OXFORD READINGS IN PHILOSOPHY

MEANING AND REFERENCE

Also published in this series

The Concept of Evidence, edited by Peter Achinstein
The Problem of Evil, edited by Marilyn McCord Adams and Robert Merrihew Adams
The Philosophy of Artificial Intelligence, edited by Margaret A. Boden
Perceptual Knowledge, edited by Jonathan Dancy
The Philosophy of Law, edited by Ronald M. Dworkin
Theories of Ethics, edited by Philippa Foot
The Philosophy of History, edited by Patrick Gardiner
The Philosophy of Mind, edited by Jonathan Glover
Scientific Revolutions, edited by Ian Hacking
Hegel, edited by Michael Inwood
Conditionals, edited by Frank Jackson
The Philosophy of Time, edited by Robin Le Poidevin and Murray MacBeath
The Philosophy of Religion, edited by Basil Mitchell
A Priori Knowledge, edited by Paul K. Moser
The Philosophy of Education, edited by R. S. Peters
Political Philosophy, edited by Anthony Quinton
The Philosophy of Social Explanation, edited by Alan Ryan
Propositions and Attitudes, edited by Nathan Salmon and Scott Soames
Consequentialism and its Critics, edited by Samuel Scheffler
Applied Ethics, edited by Peter Singer
Causation, edited by Ernest Sosa and Michael Tooley
Locke on Human Understanding, edited by I. C. Tipton
Theories of Rights, edited by Jeremy Waldron
Free Will, edited by Gary Watson
The Philosophy of Action, edited by Alan R. White
Leibniz: Metaphysics and Philosophy of Science, edited by R. S. Woolhouse
Demonstratives, edited by Palle Yourgrau

Other volumes are in preparation

MEANING AND REFERENCE

edited by

A. W. MOORE

OXFORD UNIVERSITY PRESS
1993

Oxford University Press, Walton Street, Oxford OX2 6DP
Oxford New York Toronto
Delhi Bombay Calcutta Madras Karachi
Kuala Lumpur Singapore Hong Kong Tokyo
Nairobi Dar es Salaam Cape Town
Melbourne Auckland Madrid

and associated companies in
Berlin Ibadan

Oxford is a trade mark of Oxford University Press

Published in the United States
by Oxford University Press Inc., New York

British Library Cataloguing in Publication Data
Data available

Library of Congress Cataloging in Publication Data
Meaning and reference/edited by A.W. Moore.
p. cm. — (Oxford readings in philosophy)
Includes bibliographical references and index.
1. Meaning (Philosophy) 2. Reference (Philosophy) 3. Language and languages—Philosophy. I. Moore, A. W., 1956– . II. Series.
B105.M4M39 1993 121'.68—dc20 92-27073

ISBN 0–19–875124–9
ISBN 0–19–875125–7 (Pbk.)

1 3 5 7 9 10 8 6 4 2

Typeset by Best-set Typesetter Ltd., Hong Kong

Printed in Great Britain
on acid-free paper by
Bookcraft (Bath) Ltd., Midsomer Norton, Avon

CONTENTS

INTRODUCTION

A. W. MOORE

I

Gottlob Frege, in Essay I, sets the agenda for this volume. In the course of sketching some of the main features of his seminal and wide-ranging theory of meaning, he draws an important distinction between "sense" and "reference". Though this distinction is anticipated elsewhere, Frege is the first person to articulate it so clearly and so forcefully. Later, in Essay II, he gives a more succinct account of it. Thereafter, each essay in the volume is a more or less direct attempt to develop, apply, clarify, amend, or reject it.[1]

Frege introduces the distinction in terms of what he calls *names*. These are the singular noun-phrases that are used to refer to particular things. (The term 'things' is to be understood in the broadest possible way. Examples of names are 'Plato', 'the positive square root of 16', 'the number of symphonies written by Schumann', and 'joy'.) By a name's "reference" he means whichever thing the name is used to refer to. Thus the reference of 'the positive square root of 16' is the number 4; so too is the reference of 'the number of symphonies written by Schumann', Schumann having written (in fact) four symphonies. When someone knows what the reference of a name is, they know as much about the name as is directly relevant to the truth or falsity of any declarative sentence in which the name occurs.

Even so, Frege argues, knowing what the reference of a name is is not the same as understanding it; there is more[2] to meaning than that. For example, someone who has it on good authority that the reference of a particular name in Urdu is the number 4 does not thereby know what the name means. Again, someone who knows perfectly well what 'the positive square root of 16' means, and what 'the number of symphonies written by Schumann' means, cannot tell, without further knowledge, that they have the same reference. It is in cases of this kind that sentences of the form '*a* is the same thing as *b*' are both true and informative.

[1] In the case of Essay XIV the immediate concern is exegesis of Wittgenstein. But Wittgenstein himself is engaged, very indirectly, with Frege's idea (see further below, Section X). In some of the essays the distinction itself is retained but its first member is rejected.

[2] Less? (See further, Essay XIII, pp. 243ff.)

This is where Frege introduces "senses". A name's sense is said to be what we *grasp* when we understand it. The name is said to *express* its sense. And the sense is said to *determine* its reference. Thus while two names with the same reference can have different senses, two names with the same sense (synonyms) cannot have different references.

But there is more to the claim that sense determines reference than that. A name's sense, Frege says, contains the "mode of presentation" of its reference. In grasping the sense, we think of the reference in a certain way. As a corollary, the sense fixes what the name would have been used to refer to in any other possible circumstances (so long as it had had the same sense). If Schumann had written five symphonies, 'the number of symphonies written by Schumann' would have been used to refer to the number 5.

The whole apparatus can then be extended to parts of speech other than names. It is admittedly somewhat strained to say, of some of these, that they are used to "refer" to anything, or to suggest that they have extra-linguistic correlates. (Consider a connective like 'and'.) Still, each of them does have a feature that is directly relevant to the truth or falsity of any declarative sentence in which it occurs. That provides enough of an analogy to sustain talk of reference; and, therewith, talk of sense.

But what kind of things are senses? Frege is adamant that they are not privately associated subjective ideas. If they were, ordinary linguistic communication would be impossible. For it is only when two people grasp the same sense, and know that they do, that they are able to understand a word in the same way and thereby use it to convey their thoughts to each other.

To sum up then: meaning, or more specifically sense,[3] is an objective feature of how words are used and understood; and this is not something exhausted by any simple word/world relationship.

II

Bertrand Russell is sceptical about Frege's introduction of senses, at least in the primary case of what Frege calls names. (The discussion in this section will not extend to other parts of speech.) Russell shares Frege's interest in names. But, unlike Frege, he insists that they form a very heterogeneous class. It is only by acquiescing in surface features of

[3] There is more to meaning than sense. For example, it is a feature of a word's meaning, but not of its sense, that it is a swear-word. (Many swear-words share their senses with non-taboo medical terms.)

language that we are led to assimilate them. If we use techniques of analysis that Frege himself made available, we can separate them into fundamentally different categories. Many of the problems that Frege was addressing then disappear and many of his conclusions look unwarranted. Or so Russell suggests in Essay III.

Some Fregean names, the ones that Russell himself is prepared to call names—we can call them *logically proper names*[4]—are simple indefinable words whose meanings do not depend on the meanings of any constituent expressions. In their case there is no obstacle to an identification of meaning with reference, even if there is room for controversy about which expressions fit the bill. (Russell himself discounts proper nouns such as 'Plato', which he thinks can be defined. The definition of 'Plato' would presumably be something like: 'the author of *The Republic*'.) The obstacles arise with definite descriptions—singular noun-phrases of the form 'the so-and-so'. But these function very differently from logically proper names. To see why, consider a definite description which is not in fact satisfied by anything, say 'the planet between Mercury and the Sun'. In Fregean terms, this is a name with a sense but no reference. (Frege himself takes it to be a defect of natural language that such names can be constructed.) Now suppose we understand it as *purporting* to have a reference in the same way in which a logically proper name does. What then are we to say about a declarative sentence such as 'The planet between Mercury and the Sun is green'? We are forced to say that, since it is not about anything in the way in which it purports to be, it is incapable of being either true or false, and is hence meaningless. But that, for Russell, is absurd. He thinks the sentence is perfectly meaningful (and not because it is somehow about an "unreal object" instead). He concludes that we must understand the description differently. We must disregard the sentence's superficial grammar, and stop thinking of the description as a proper constituent of it. Hence Russell's analysis: 'The planet between Mercury and the Sun is green' is equivalent to the conjunction of the following three sentences:

(i) There is at least one planet between Mercury and the Sun
(ii) There is at most one planet between Mercury and the Sun
(iii) Any planet between Mercury and the Sun is green.

[4] This is what Russell is often said to have called them. (For example, Strawson attributes this phrase to him in Essay IV, p. 59). But I have not been able to find the phrase anywhere in Russell's writings. (I discount 'Mr. Strawson on Referring', in *Mind*, 66 (1957), 386, where Russell himself is quoting Strawson.)

Given that there are no planets between Mercury and the Sun, this conjunction is straightforwardly false—because of (i).

What are the implications for sense? Well, those Fregean names for which an identification of meaning with reference is ruled out (definite descriptions) are not, on this conception, proper constituents of sentences at all, with any kind of meaning of their own. It is a mistake even to ascribe references to them. And a sentence of the form 'The so-and-so is the same as the such-and-such' can be both true and informative, not because there are two different modes of presentation of the same thing but simply because it is both true and informative that at least one thing satisfies both descriptions (or, for that matter, either one of them). The arguments for sense simply lapse.

III

There is a counter to the Russellian analysis in Essay IV, by P. F. Strawson. Strawson develops some thoughts that Frege himself has in Essay I, but in a way that is unlikely to satisfy Frege, since Strawson shows greater respect for the workings of natural language. The result is a more restrained, less theoretically committed view of meaning and word use.

First, Strawson argues, we must stop trying to locate meaning in some invariant relation in which words and phrases stand to the world. For that is to ignore the important role that context has to play, together with meaning, in determining what we are talking about. Thus consider the rather free use that has so far been made of the notion of "the reference" of a singular noun-phrase. To talk in those terms is to blur the important fact that singular noun-phrases can be used to refer to *different* things in different contexts, as witness a demonstrative pronoun like 'that'. Both Frege and Russell are aware of this fact. But it does seem to invalidate Russell's principal argument. For it points up an important distinction which the argument overlooks. There are declarative sentences. And there are particular utterances of them. Meaning attaches to the sentences, truth and falsity to the utterances—or at least, to assertions made by means of those utterances. The two things are separate. It is perfectly possible for a meaningful declarative sentence which can be uttered in a suitable context in making an assertion that is true, and in another suitable context in making an assertion that is false, to be uttered in an *un*suitable context in failing to make an assertion that is either true or false. The last is what happens when someone says, 'That is green,' without succeeding in indicating anything. Likewise when someone says, 'The thing on the table is green,' when nothing is on the table. And likewise, presumably, when someone says, 'The planet between Mercury and the Sun is green' (though in this case there appears to be no such thing as a suitable context). There now seems to be no reason why we should not assimilate definite descriptions to other Fregean names, in the way in which Frege himself does. We simply have to be careful to relativize our ascriptions of reference to them. Thus in one context the reference of a description may be one thing, in another another, and in yet another—why not?—the description may have no reference at all.

Russell, as we have seen, takes a different view of the sentence 'The planet between Mercury and the Sun is green.' He maintains that any utterance of this sentence is false. So he is led to identify as part of the meaning of the sentence, and part of what is asserted, something which for Strawson is merely "implied" by the utterance about the suitability of the context—that there is at least one planet between Mercury and the Sun.

It is hard to know how to arbitrate between these two accounts. Strawson's seems less barbarous to untutored intuition. But how do we tell? And so what? Would the fact that people are disinclined to call certain utterances false be any more significant than the fact that people are disinclined to call glass a liquid?

Russell's analysis has a certain tidiness and elegance. It gives an especially crisp account of why some sentences are ambiguous, for example 'This time yesterday the thing on the table was green.' There are two ways of subjecting this sentence to Russellian analysis: either 'this time yesterday' governs all three of the conjuncts, and the sentence is concerned with what was then on the table; or 'this time yesterday' is part of the inner structure of the third conjunct, and the sentence is concerned with how the thing on the table then was. But is such analysis too artificial to cope adequately with the perhaps untidy, inelegant logic of everyday discourse?

There is a further question to consider. In so far as we are concerned with the "logical form" of certain sentences, is there any reason to suppose that each of them has (in this respect) a unique logical form? Might there be some contexts whose logic made it inappropriate to treat a given sentence as if it involved a logically proper name, and other contexts whose logic made it inappropriate to do anything else?

IV

There is an attack on senses of a quite different and very radical kind in the work of W. V. Quine. Quine takes meaning to be ultimately a matter of observable behaviour and dispositions to behaviour on the part of language users. There is a point here that is more or less common currency: a word's meaning can never be discerned except by observing how it is actually used. (Even an onomatopoeic word has to be used as such.) But the point is given a distinctive twist by Quine, who derives it from a general physicalism. The only facts are physical facts (facts about how things are physically). Quine believes that this is enough to discredit talk about senses, and any implicated talk. The connection is not obvious. It is not obvious that the physical facts of language use cannot justify the attribution of different senses to different expressions. But Quine has an argument designed to show that they cannot.

The argument runs as follows. The notion of a physical fact is bound up with physics. Physics is at the centre of our general picture of the world. It is the result of our best attempts to organize and systematize what we know about the world through observation. But it is not dictated to us. Observation does not rule out alternative and equally good systems of the world. This means that, when we turn to our linguistic behaviour—in particular, to the way in which we use the language of physics, in its more theoretical reaches—there are different, equally good ways of locating and assigning meaning. For our linguistic behaviour is governed not only by meaning but also by our view of things. However, given the slack just referred to, it could just as well be described as if it were governed, with different meaning, by one of the rival views. Thus suppose we respond to the evidence that we have amassed by endorsing some scientific formula. The physical facts of the case (our observations and our subsequent verdict) admit of rival interpretations. It may be that we have come to the view that nature is governed by law L, and our formula means precisely that; or it may be that we have come to the view that nature is governed by law L^* (quite distinct from L), and our formula means precisely *that*. Since these different interpretations are not themselves part of physics, nor, as we have seen, constrained by physics, there is (literally) no fact of the matter concerning which is right. Meaning is indeterminate. To assume senses, however, is to assume determinacy.

Why does Quine think that to assume senses is to assume determinacy? There are two (connected) reasons. The first is that senses attach to individual expressions. But the indeterminacy of meaning arises, on Quine's view, because there are no facts about meaning beyond behaviour; and behaviour reaches to language as a whole, allowing variation precisely in the distribution of meaning over its parts. If our formula is interpreted in one way rather than another, then related formulae, including formulae we reject, must be interpreted in a complementary way.

This is not decisive, however. References attach to individual expressions too, yet Quine does not have the same objection to them. Why not? Because he thinks that a particular assignment of references is one legitimate way among others of carving up the facts of linguistic behaviour. It can amount to a statement of those facts. An assignment of senses, on the other hand, introduces further resolution and adverts to further facts—but without further warrant.

Quine's second reason for believing in the connection between senses and determinacy is that senses are purely semantic and independent of any world-view. They fix, by themselves, what the references of expressions would be in any possible circumstances. But (again) the indeterminacy of meaning arises, on Quine's view, because there are no facts about meaning beyond behaviour; and meaning governs linguistic behaviour only inextricably with some (empirically underdetermined) world-view. Suppose that a system of the world has to be supplemented, or perhaps modified, in the face of some new experience. Then, if Quine is right, there will be quite different ways in which our linguistic behaviour can express the supplement, or the modification, any one of which can count as meaning-preserving. In particular, it is left open which new sentences are to be affirmed. This means that it is always idle to try to reach conclusions *via* questions of the form, 'What would we say if such and such?' There are no facts of the matter.

Quine has explored these ideas through all their various consequences. Thus, to take a well-known example, there is a distinction that arises if each declarative sentence has its own sense. Some true sentences[5] are then "analytic"—true just *because* of their senses. (To understand an analytic truth is already to be able to see that it is true. Its truth extends to all possible circumstances. A putative example is 'All vixens are female.') The rest are "synthetic"—true also, in part, because of how things are. (To understand a synthetic truth is *not* yet to be able to see that it is true. Its truth depends on the actual circumstances. A putative example is 'All vixens are less than five feet in length.') For reasons

[5] Attention is restricted to those sentences (if such there be) which can be classifed as true or false without reference to individual utterances of them.

sketched above, there is no behavioural way of substantiating this distinction, just as there is no saying, determinately, whether the rejection of a sentence previously held true represents a change of view or a change of meaning. Quine does allow a distinction of degree. Thus he pictures sentences as forming a fabric, each lying more or less close to the edge where the fabric "impinges on experience".[6] That is, the truth or falsity of each is more or less sensitive to how experience reveals things to be. But even to talk of an edge may be to concede greater determinacy than his own arguments allow. (As an aside: can Quine perhaps admit a distinction at the level not of sentences but of utterances? We frequently introduce definitions into our conversations as a way of imposing clarity and aiding discussion. There seems to be no harm in Quine's allowing that in a limited context of that kind the utterances of certain sentences will bear the mark of analyticity.[7])

Midway through Essay V, in the course of outlining these ideas, Quine turns his attention to translation. This offers a new and perhaps sharper focus. Any indeterminacy of meaning must issue in a corresponding indeterminacy of translation. For if meaning is indeterminate, there can be different, equally good ways of compiling a bilingual dictionary—at the limit, "different" to the point of being incompatible, and "equally good" to the point of being each perfect. Conversely, given senses, any perfect bilingual dictionary must respect them, and any incompatibility between two dictionaries will show that at least one contains an error. If both dictionaries satisfy all the physical constraints that can be imposed on their adequacy, still there is a fact of the matter concerning which, if either, is right. But note: this is an issue about how we should *assess* a situation in which there are two competing dictionaries of this kind. It is not an issue about how likely such a situation is to arise. There are no new questions of (empirical) substance here.

There is, however, the following new consideration. If, with Quine, we think that there can be two perfect but incompatible dictionaries, then we owe an account of what we mean by 'incompatible' here. For obviously we cannot mean 'differing at some point as to sense'. On the other hand, other differences of meaning are not, by his own lights, differences in *fact*. It looks as if the Quinean position is self-stultifying. (This is a charge that has often been levelled against him. There is similar cause for concern in the idea of alternative and equally good systems of the world. In what way "alternative"?)

[6] This phrase occurs in his 'Two Dogmas of Empiricism', reprinted in *From a Logical Point of View* (New York: Harper & Row, 1961), 42.
[7] Cf. what he says in ibid. 35 (though his point there is different).

Quine has a ready reply: incompatibility (like alternativeness) is a matter of details seen out of context. Two dictionaries are incompatible if they contain elements that could not consistently be combined into a single dictionary (or, more cautiously, if there is a particular translation which is correct according to one dictionary and incorrect according to the other). For example, it may be that there is, in one of the two languages involved, a scientific formula which is equivocal in the way considered earlier and which one of the dictionaries renders one way, the other the other. Each dictionary, taken separately, can still fit the relevant patterns of overall behaviour perfectly. This will show up most clearly if the "target" language—the language into which a translation is to be made—allows for finer and subtler discriminations than the other, and affords correspondingly wider scope for reproducing those patterns. (But the discrepancy had better not be too great, lest the fit be imperfect.)

The upshot is a powerful attack on senses. There can be no legitimate statements of meaning that are not couched ultimately, albeit indeterminately, in terms of the simple word/world relationship.

V

Suppose Quine's scruples are justified, and his conception of meaning broadly correct. Donald Davidson has done as much as anyone to show what can still be achieved in the theory of meaning.

Davidson thinks that we shall have said all that we fundamentally need to say about meaning, and in particular we shall have provided for an account of understanding, once we have said how this word/world relationship is to be specified. But we must do so without at any point taking for granted thicker semantic concepts (above all, of course, the concept of sense). This is the essence of Davidson's project. Central to the project is truth. What is required, Davidson argues, is an account of what a *theory of truth* would look like for a given language, that is, roughly, a formal, axiomatic theory which states when[8] any given sentence of the language, or an utterance of it, is true. (There is nothing particularly recondite about this. An utterance of 'That is green' is true when the thing indicated by the speaker at the time is green. The centrality of truth to the project connects with the broad characterization

[8] 'When' must be understood in a suitably innocuous sense. It must not mean 'in which possible circumstances'. But it can mean 'in which (actual) circumstances'.

of the word/world relation proffered in Section I: according to that characterization, what it is for something to be the reference of an expression is a matter of what makes sentences, or utterances of them, true.)

The fundamentals of Davidson's project are set down in Essay VI. It has a Quinean underpinning in that truth itself is a function, albeit indeterminate, of which utterances speakers of the language hold true (a connection that Davidson elaborates elsewhere); and which utterances speakers of the language hold true can be determined from their observable behaviour. A theory of truth for a given language is an empirical theory. But it is also, or at least it can serve as, a *theory of meaning* for the language. Loosely, this means that it "gives the meaning" of every sentence in the language. Less loosely, it means that, if anyone knew the theory, they would *ipso facto* understand the language.[9]

But there is a question about how anyone *could* know the theory. More particularly, how could they know it given that it extends to indefinitely many sentences, most of which they have never encountered and never will? We had better do something towards answering this question if we are to provide for an account of understanding, as intended.

In fact, of course, a person's understanding of a language is based on familiarity with the individual words that comprise its finite vocabulary. (That the meanings of sentences depend on the meanings of words is indeed Davidson's starting-point in Essay VI.) A theory of meaning for the language should reflect this fact. It should have finitely many axioms, a separate one for each word and each construction in the language. And this will explain how knowledge of the theory, by beings limited as we are, is possible.

The axiom governing any given word will state, in effect, what its Fregean reference is. Davidson is addressing the word/world relationship directly, unlike Quine, in his discussion of translation, where he focuses on a word/word counterpart. Davidson has no alternative. He is interested in the kind of theory that could confer understanding. So he has to look to something that says, straight off, how words relate to the world. (Possessing all the information in a French/English dictionary could not confer understanding of either French or English on a

[9] To be even less loose, as Davidson has subsequently emphasized ('Reply to Foster', reprinted in his *Inquiries into Truth and Interpretation* (Oxford: Clarendon Press, 1984), 174), we should add that they must know the theory *as* such a theory.

monolingual speaker of German.) Still, this does help to emphasize that any theory which satisfies his requirements can be non-vacuously stated in the very language with which it deals: the purely contingent semantic relationship that holds between the English word 'green' and green things is none the less contingent for being specified in English. So, since we can be confident that such specifications are correct, without putting in any special empirical effort, it is a trivial matter, so far as that goes, to give examples of what the axioms and theorems of a theory of meaning for a language should look like.[10]

In some respects, then, it is easier to work directly with the word/world relationship than to work with its word/word counterpart. In other respects, as Davidson makes clear, it is less easy. 'Si' means the same as 'if'. But what does 'if' mean? Or: what is its reference? Or: how is it semantically related to the world? (Knowing how to frame the question is half the problem.) It is of little avail to know that an answer can incorporate the word 'if'. 'The reference of "if" is if' is gibberish. Saying how to specify the word/world relationship is a *non*-trivial matter that raises deep questions about how sentences are structured—questions about how meaning, truth, and reference ultimately come together. Hence much of the interest in Davidson's work.

But one crucial matter has been slurred over. Is Davidson claiming that when someone is fluent in a given language they are thereby *conversant* with a theory of the kind he describes (if only implicitly)? Or is he claiming that being conversant with such a theory would be just one way of understanding the language (a way that may not be available to us)?[11] If the former, how do we tell? If the latter, then surely we still need an account of what form (our) understanding actually takes; and the insistence that such a theory should have finitely many axioms may lose its rationale. Either way, do we not need to say what being conversant with such a theory would *be*? And how confident are we that we can do that without reintroducing Fregean senses? (Remember how Frege himself introduced them.) The Quinean revolt is in danger of subsiding into Fregean conservatism.

[10] Or so Davidson claims. But perhaps we should be suspicious about this easy combination of confidence and contingency. They are not incompatible. Even so, it is worth asking whether we have exaggerated the extent to which using a word twice, once with inverted commas round it, manages to display some distance between the word and the world in a way that can lead to serious theorizing about meaning. (This is a bit of axe-grinding. See my 'How Significant is the Use/Mention Distinction?', *Analysis*, 46 (1986), 173–9.)

[11] What he actually says, p. 100, is that such a theory should "recover the structure" of understanding.

VI

One possible response to this challenge, the response given by John McDowell in Essay VII, is to yield to it, but willingly. McDowell thinks that Fregean conservatism can now be made to look much more acceptable than it did. Senses had begun to look like mysterious intermediaries between our language and the world, cut loose from all empirical data. But Davidson's work can be taken to show that they are nothing of the sort; and that we can talk about sense without deviating any further from authorized plain speech than when we talk about reference. Reconsider what a theory of meaning for a language is. It is a theory such that, if anyone knew it, they would *ipso facto* understand the language.[12] But senses are simply what are grasped in understanding. So there is nothing more to be said about what the sense of an expression is beyond what somebody would know about the expression in knowing a theory of meaning for its language. Thus consider Frege's own example of two names with the same reference but different senses—'Aphla' and 'Ateb'—as set down in Essay II. The difference between their senses is simply the difference between knowing that the reference of 'Aphla' is Aphla and knowing that the reference of 'Ateb' is Ateb. (Frege's original point was that one could know both of these things without realizing that 'Aphla' and 'Ateb' had the same reference. For one may not realize that Aphla and Ateb were the same thing.) The simple expedient of recycling the very words under discussion, which earlier gave us confidence in our handling of reference, now looks set to give us similar confidence in our handling of sense.[13]

The question is: how far does this address the worries expressed at the end of the last section? All that McDowell has done, on a hostile view, is to accept something considerably more anaemic than Frege's own conception by not addressing those worries at all. For instance, concerning the question of whether knowing a theory of meaning for a given language is both sufficient and necessary for understanding the language, or only sufficient, McDowell insists that it is only sufficient, but denies that this means shirking problems about understanding our own understanding. Again, he denies that we must burrow beneath a theory of meaning to say what being conversant with it would *be*: everything that we need to know in this regard is open to view in the theory itself as it systematically relates words to the world. McDowell is quite self-

[12] But see above, n. 9.
[13] But see my reservations voiced above, n. 10.

conscious about this apparent evasiveness (just as he is self-conscious about the fact that his own conception of sense lacks some of the richness of Frege's). He argues that both the urge to dig deeper, and the fear that all will not be well if we do, arise from a misguided conviction that we must uproot the psychological mechanisms which underlie our linguistic behaviour. We feel we must explain how we manage to advance to such behaviour from our original ignorance of the language. This means characterizing the behaviour in a way that presupposes nothing about meaning. For McDowell, on the other hand, the important thing is to learn to accept the linguistic surface for what it is, without thinking that we either can or need to account for it in more fundamental terms. We shall come back to this point in Section X.

Meanwhile, however, the worries linger. It seems that we *must* explain why, if Aphla is Ateb, there is nevertheless a difference between knowing that the reference of a name is Aphla and knowing that its reference is Ateb. This is precisely the kind of thing which Quine has argued we cannot do without an illicit concession to determinacy. The thought persists that we may be close once again to an extreme Fregean conservatism.

VII

This thought has been pressed most notably, and most urgently, by Michael Dummett. In Essay XIII he puts it directly. He argues that if Davidson's work is to form the basis of a satisfactory account of understanding, then it must, after all, be supplemented by something more like Frege's own original sense/reference distinction. Furthermore, until we have a satisfactory account of understanding, we do not have a satisfactory account of meaning. Meaning and understanding are correlative.

Dummett's emphasis on understanding involves him in the same belief in the importance of behaviour—of the actual use of expressions by those who show understanding—that motivates Quine and Davidson. But in Essay VIII he explains why it leads him in a different direction from them, just as it leads him in a different direction from Wittgenstein, and from others who have been likewise motivated. The common ground among these thinkers can be put in the form of a rough slogan: meaning is use. But there are myriad different ways of setting off from there. Dummett, like McDowell, is reluctant to set off in the direction of reductionism. Reductionism, in the sense in which I intend it—

there are other, more demanding senses—is the view that use can be conceived quite independently of meaning, that talk of meaning is just another way of representing facts about use. Thus Quine. Dummett is equally reluctant, however, to set off in the opposite direction, with Wittgenstein (and, to a lesser extent, McDowell himself). That way lies the view that use is so much determined by meaning, and meaning by use, that not only must meaning be grasped before use itself can properly be conceived, but there is no prospect of singling out from the facts of use, and subjecting to systematic theoretical enquiry, what is thereby grasped.[14] For Dummett, what is important is that meaning and understanding should be open to public view. Otherwise nobody could ever learn what an expression meant or demonstrate that they had done so, and language itself would be impossible. But the governing view of language remains broadly Fregean. In particular it assigns a distinctive role to truth, as it relates to the sense/reference distinction.

There is an important new twist though. Towards the end of his essay Dummett develops the governing view, as he has in a number of other places, to cast doubt on the particular conception of truth that Frege himself adopts. And this is not just an *ad hominem* point. The conception in question is thoroughly orthodox. It supports some of our deepest metaphysical convictions about the world.

Dummett's concern is with the idea that every statement (assertion, utterance of a declarative sentence, or what you will) is true or false. He thinks that there are some superficial reasons for challenging this idea and some very deep ones. The superficial reasons are of the kind that Strawson canvasses in Essay IV. If they hold, they give us a way of recognizing certain statements as neither true nor false (because we can see that some Fregean name lacks a reference, say).[15] They are superficial because they do nothing to threaten the idea that every statement is either true *or not*. Let us therefore put them to one side by agreeing, henceforth, to call a statement 'false' when it is not true. The deeper reasons persist. These have to do with what we do *not* have any way of recognizing. Suppose, for example, that we shall never be able to tell whether a given statement is true or false—that we have no way of recognizing which of the two it is. (Consider a statement involving speculation about some distant and undocumented event in history.) Then, the thought is, we have no justification for assuming that it *is* one or the other. To insist otherwise is precisely to credit the statement with

[14] McDowell's commitment to Davidson's project prevents him from going all the way down this path.

[15] It is true that we might refuse to dignify such statements with that label: we might prefer to call them 'pseudo-statements'. But that is an unimportant point of terminology.

a meaning that goes beyond what can ever be open to view in our use of language, for we shall never be in a position either to affirm the statement or to reject it.

It might be protested that this involves a radical departure from how we normally think and speak which Dummett himself should be the last to sanction, given his own call to respect our actual linguistic practices. Dummett's reply is that, unlike Quine, he is not a holist about meaning: our linguistic practices are not simply "given" *en bloc*. (In Essay XIII he touches on the reasons why not.) Some of those practices can be criticized for not harmonizing with others.

A different objection is that he cannot strictly say in what circumstances the assumption that an arbitrary statement was either true or false would be unfounded. If we could never tell which of the two it was, then that would itself be something which—on this "anti-realist" understanding—we could never tell. (For if we could, then we could tell that we could never tell that it was true: but to do that properly would be to tell that it was false.) To insist that such circumstances could arise is therefore itself to credit statements with meanings that go beyond what can ever be open to view in our use of language. It seems to me that this objection, even if Dummett has an answer to it, cuts very deep.

VIII

A further revolt against Frege comes in the work of Hilary Putnam and Saul Kripke, represented in Essays IX and X. They revert to a Russellian view of names. But their conception of names is not as narrow as Russell's—nor, importantly, as broad as Frege's.[16] They are concerned with what are often called 'proper names': 'Plato', 'London', 'Venus', and the like. And they extend the view to "natural kind terms" such as 'water', 'cat', and 'human'. Let us call any expression which their view embraces a *Name*.

They argue that it is impossible to explain the semantics of a Name in terms of Frege's sense/reference distinction. There is nothing which plays the role of sense. There is nothing which (i) we grasp when we understand the Name, which also (ii) determines its reference. The things which satisfy these conditions are different and separable (and the Name's meaning is best identified, *à la* Russell, with its reference). What we grasp when we understand the Name—what satisfies (i)—may be

[16] Russell, remember, discounted proper nouns such as 'Plato'. He thought that these could be defined and were thus disguised definite descriptions. Frege, on the other hand, counted even undisguised definite descriptions.

public, objective, and communicable; but it is liable to vary from one speaker to the next, and the actual grasping of it is a matter of individual psychology. (Not that Frege denied this.) What determines the reference of the Name—what satisfies (ii)—is something altogether different, something which is not encapsulated in any individual view of things. It works as follows. There is an initial "dubbing", whereby the Name is directly associated with some thing or kind. Thereafter, for the meaning of the Name to be preserved is, quite simply, for it to be used to refer to the same thing or same kind. And this is guaranteed if each use lies on an appropriate causal route from the initial dubbing. ('Appropriate' will come under scrutiny in the next section. But roughly, the route must consist of a series of uses of the Name, each of which—bar the first—involves the intention to preserve the meaning of its predecessor.) Speakers may have their own different ways of telling whether any given thing *is* the thing in question, or of the kind in question; or they may defer to experts; or they may acquiesce in their ignorance. But that is irrelevant to whether they successfully carry out their intention to use the Name in a meaning-preserving way. The world itself is now seen playing a more dominant role in the word/world relationship. The same narrowly described patterns of word use and individual psychology, on one side of the relationship, could theoretically accompany quite different things on the other side—if they derived from different dubbings, which secured different references. (The different references might be two kinds of stuff that were superficially indistinguishable, gold and fool's gold for example.) Hence the split between (i) and (ii).

But how radical an attack on Frege is this? Many of the same Fregean pieces are still in play—only rearranged. For instance there is still something being said to satisfy (ii). A direct determination of a Name's reference, achieved by a dubbing rather than by what anyone grasps when they aim to use the Name correctly, is, none the less, a determination. From a purely semantic point of view, all that matters about a Name, and all that the dubbing needs to secure, is its reference. (This is why its meaning and its reference can be identified.) But the reference is determined for all that. In fact it is determined in the most robust sense: there is a fact about what the Name would have been used to refer to in any other possible circumstances in which the reference had been determined in that way. It is just that, because the determination is direct, the Name would have been used to refer to whatever it is used to refer to in the *actual* circumstances—provided that the thing in question, or the kind in question, had still existed. This is what is meant by calling the Name a 'rigid designator'.

That reference *is* thus determined in this account is clear from the vital role that thought-experiments play in the arguments for the account. These thought-experiments are supplemented, in Putnam's hands, by appeal to an imaginary planet which he calls 'Twin Earth', and, in the hands of both Putnam and Kripke, by appeal to different "possible worlds" (though there are qualifications about these *not* being like different planets). The use of such devices is simply to tease out what it would be right to say in different possible circumstances, and to explore what could or could not have been the case. The whole point of Kripke's essay, and a cardinal point of Putnam's, is to deliver verdicts on what is necessary and what contingent. (But the (i)/(ii) split introduces a parallel split here. A proposition can be *metaphysically* necessary—necessary with respect to what satisfies (ii)—yet *epistemically* contingent, or, as Kripke says, a posteriori—contingent with respect to what satisfies (i). The most familiar example is the proposition that water is H_2O. The reference of 'water' is determined in such a way that this *must* be true: water *is* H_2O. But we, reflecting only on what we grasp in our casual understanding of the term, can envisage other possibilities.)

Putnam's and Kripke's account is broadly Fregean then. Indeed had they operated with tighter criteria for a full and proper understanding of a Name—in particular, had they insisted that nothing less than a grasp of what satisfies (ii) should count as such—then the notion of sense would not have come apart in the way that it did, and they could have spelt out their account as follows, with a straightforward appeal to the sense/reference distinction.

> Each Name has a sense and a reference. But the sense is of a special kind, such as to ensure that the Name would have had the same reference in any possible circumstances in which that reference had existed. Not everyone who correctly uses the Name grasps this sense. Maybe only a few experts do, or maybe things have been so contrived that no one does. But that is to say that not everyone who correctly uses the Name fully and properly understands it. The sense of the Name contributes to the analyticity of certain sentences in which it occurs.

(These analytic sentences correspond to Putnam's and Kripke's "metaphysical necessities". *They* classify them as synthetic. But this is just another reflection of the original (i)/(ii) split. Such sentences are analytic with respect to what satisfies (ii), synthetic with respect to what satisfies (i). Putnam and Kripke—arbitrarily, as it seems to me—are using the terms 'analytic' and 'synthetic' with implicit relativization to what satisifes (i).)

IX

Putnam and Kripke are advocating a version of the Fregean view, then—on a suitably narrow conception of understanding. But are they doing so even apart from that conception? David Wiggins, in Essay XI, suggests that they are. He first canvasses Putnam's account and develops it in various ways. He then concludes, without at any point tightening his criteria for understanding, that what Putnam has done (though Putnam does not himself see it in these terms) is to argue that natural kind terms have senses of a special sort.

Wiggins is not deterred by this conclusion. He thinks that natural kind terms do have senses of a special sort, and that reflection on how we actually learn them and use them amply reinforces the point. Compare this with the way in which McDowell was happy to accept the charge of Fregean conservativism in Essay VII. The resultant position can be compared with McDowell's in another way. Wiggins thinks that one conveys the sense of a natural kind term by conveying its reference. The sense is not some intermediary between the term and its reference, but rather a particular *conception* of its reference. He ends his essay by adverting to some of the possible ways of explaining what this means.

This Fregean reading of Putnam's and Kripke's account is given further support by a critique of Kripke's proposals which Gareth Evans offers in Essay XII,[17] and which Dummett takes up in Essay XIII.

Evans argues that Kripke's proposals must first of all be amended and refined. The causal routes from a dubbing cannot *just* involve uses of the Name intended to preserve meaning. If they did, there could not be unintended changes of meaning. Yet these are perfectly possible, as Evans makes clear. He cites the example of 'Madagascar', originally the name of part of the African mainland but erroneously transferred to the island. Kripke's account threatens to have us referring still to the mainland when we use the name. To block this kind of absurdity, we must go back to what individuals grasp when they use a Name. In particular we must look to the body of beliefs and information which they associate with it. These do not determine the Name's reference in the way in which Frege's account was thought to imply. For it is not a question of what *fits* them. (They may be mostly incorrect.) But they are relevant. It is a question of what their causal source is. Kripke is therefore quite right to insist that what determines the reference of a

[17] The particular proposals under scrutiny are in fact those which Kripke spells out in a related essay, *Naming and Necessity* (Oxford: Blackwell, 1980).

Name involves a causal route leading up to current uses of the Name. But he specifies the wrong route. So a correct account of reference, according to Evans, lies somewhere between what Kripke offers and what he wants to reject.

But then Dummett pushes the same idea further. He argues, just as Wiggins did, that once this has been properly thought through, we see that the correct account is indeed an orthodox Fregean one. For one thing, the whole discussion has to be limited to a much narrower range of Names than has been claimed. In the case of, say, 'Islam', or 'Hindi', any supplement to Frege along the lines of Putnam and Kripke is plainly inapplicable. But more importantly, even for those Names about which this amended causal story can be told, it seems clear, by the time the details have been spelt out, that what is being offered is not something that dispenses with Fregean senses, but, once again, a specific proposal about what those senses are like. It *is* what we grasp when we understand the Name—concerning the origin of the information that we associate with it—which determines the Name's reference: (i) and (ii) come back together.

If it is true that what Putnam and Kripke are proposing is not ultimately an attack on Frege, then *a fortiori* it is not of a piece with the radical attack which we saw in Quine (as indeed is evidenced by its commitment to notions of analyticity and necessity). But this is significant. It means that what Putnam and Kripke are proposing *still has Quine's attack to contend with*. Consider: how far is it a language-independent fact, in the way that seems to be implied, that there are certain things or kinds with which Names can be associated by means of dubbings? Will not different systems of the world recognize different things, and different kinds, and will this not already affect the use of the language in which they are couched? (Is joy something that can be dubbed? Do women and girls form a natural kind? How do we tell?) Again, suppose that we are invited to agree that we would not (now) call something 'water' if we found out that it was not H_2O. Who says? Is it a matter of linguistic intuition? What if yours differs from mine? Remember: if senses are repudiated, then speculation about how we would or would not, should or should not, use the term 'water' in different possible circumstances has nothing to answer to; and it is always idle, as I commented in Section IV, to try to reach conclusions *via* questions of the form, 'What would we say if such and such?'[18]

[18] But see Essay XI for some hints about how to respond to these challenges; see also Putnam's 'Is Semantics Possible?', in H. E. Kiefer and M. K. Munitz (eds.), *Language, Belief and Metaphysics* (Albany, NY: State University of New York Press, 1970), 50–63.

X

Dummett has risen to the Quinean challenge with as much spirit as anyone. In Essay XIII he takes each of the attacks on Frege's paradigm which we have looked at, including Quine's; pits them against what he sees as Frege's own original arguments; and concludes in favour of Frege. If we grant the background conception of reference, and if we grant that meaning is what is known or grasped in understanding, then Dummett's somewhat reconstructed Frege looks unassailable. Quine no doubt will respond with scepticism about the idea of what is known or grasped in understanding. But that just confirms that any departure from Frege's paradigm must be in the direction of something very radical indeed.

Dummett himself, however, has already pointed in such a direction in Essay VIII. This was to Wittgenstein's well-known reservations, echoed to some extent by Strawson in Essay IV, about any approach to language with theoretical pretensions of the kind we have been considering. If such reservations are justified, it is foolhardy even to talk in terms which make the sense/reference distinction an issue. For it is foolhardy to imagine that so much of theoretical interest can be contained in a statement of meaning. (We should regard a statement of meaning as the unspectacular affair that it is. One states what an expression means by citing a synonym, or by giving examples of its use, or suchlike.) It is not that Wittgenstein thinks that language and meaning are of no theoretical interest. That could scarcely be further from the truth. Rather he has a distinctive, anti-scientistic conception of how philosophical puzzles about language arise, why they matter, and how they should be addressed.

Consider, for example, a difficulty that we tend to get into when we start reflecting self-consciously on how we apply words in new circumstances. In Fregean terms, it is a difficulty concerning what is involved in the grasp of senses. In somewhat looser terms, it is a difficulty concerning what counts as carrying on "in the same way". Clearly, the concept of carrying on in the same way is integral to meaningful language-use (that is, to the recurrent use of words with constant meaning). But this seems to require that certain things should definitely count, and that others should definitely not count, as carrying on in the same way. The problem is that it is hard to see, on reflection, what can possibly sustain this, if not our simply and capriciously *taking* certain things to count and others not, which is not sustenance enough. The very idea of meaningful language-use—of meaning itself—is under threat.

Wittgenstein, by providing a more sensitive, more subtle and more sustained critique of this problem than anyone else, helps us on our way towards a natural solution. But he does not do this by arguing, in any orthodox way, for a particular view of meaning. Rather, by a careful interlacing of hints, suggestions, and platitudes, combined sometimes with mundane empirical observations, he gets us to explore the view that we already have. The problem arises because we have an ill-conceived urge to go beyond that, and to seek explanation where it is inappropriate.

What results from Wittgenstein's critique is a new sense of how far down we can dig in our investigations into language. (We saw this already when looking at McDowell's first essay in Section VI.) Bedrock is how we actually carry on, in all its variety, unpredictability, and untidiness. *But bedrock includes meaning.* And it is precisely when we try to dig deeper than that—when we start looking for what underlies meaning and what justifies our using expressions in the way that we do, when we start theorizing about language in such a way as to allow talk of a sense/reference distinction—that we get into difficulty. The hard thing is to resist the temptation to dig. We need to maintain a clear view of our carrying on simply for what it is. If we can do that, then we can at the same time attain a better understanding of the slogan that meaning is use. It is not that meaning can be reduced to use. (That was the conception which both McDowell and Dummett repudiated, according to which something in a word's use, accessible to those who do not already understand the word, serves to determine its meaning. This suggests that use lies at a deeper level than meaning.[19]) It is rather that meaning permeates use. We manage to discern meaning in one another's use of language because we have shared interests and a shared sense of what is significant and what salient. This is all brought out beautifully by McDowell in Essay XIV, where he takes up the awkward task of exegesis.

That Wittgenstein has taken so long to come into prominence in these essays, or even to receive much in the way of acknowledgement, is remarkable. He is the one person since Frege who substantially changes the terms of the debate. He points to a conception of meaning which is more profound in its threat to Fregean senses than anything we have seen hitherto. (I include Quine.) And there is considerable interest in

[19] But note that McDowell, in both his essays, alleges that Dummett himself is under the sway of this conception.

him—a good deal of it sympathetic—among these very writers.[20] Yet the discussion up to Essay XIV has kept him largely in abeyance.

The explanation for this anomaly is complex. It has something to do with Wittgenstein's own ambivalent attitude to the very practice of philosophizing about language. While he has a uniquely keen sense of its importance, he also, for reasons that we have glimpsed, deplores the kind of systematic theorizing about meaning to which so much in these essays is a contribution. Instead, he demands something more piecemeal, more descriptive, and more exploratory, together with a reappraisal of just which questions are the right ones to ask. To have reckoned with that demand in the context of these essays would no doubt have seemed like a distraction.

Whatever we make of Wittgenstein's views, I do not think that there can be any serious doubt that they should command our respect. Quite what their implications are for the dismantling of what we have been looking at is far harder to say. But it is worth remembering that Wittgenstein himself would have had a much less profound interest in language if there had not been an urge to indulge in precisely this kind of systematic theorizing, an urge which was as strong in his case as it was in anyone's.[21] Even if his views are correct, and certainly if they are incorrect, that urge should itself command our respect.[22]

[20] Much of McDowell's Essay XIV, for example, is concerned with Kripke's influential exegesis of Wittgenstein, in *Wittgenstein on Rules and Private Language* (Oxford: Basil Blackwell, 1982).

[21] I am alluding, in part, to his earlier work, encapsulated in his *Tractatus Logico-Philosophicus*, trans. D. F. Pears and B. F. McGuinness (London: Routledge & Kegan Paul, 1961), to which so much of the mature work is a reaction.

[22] I should like to thank Philip Percival, Andrew Rein, and esp. David Wiggins for their very helpful comments.

I

ON SENSE AND REFERENCE

GOTTLOB FREGE

Equality[1] gives rise to challenging questions which are not altogether easy to answer. Is it a relation? A relation between objects, or between names or signs of objects? In my *Begriffsschrift*[2] I assumed the latter. The reasons which seem to favour this are the following: $a = a$ and $a = b$ are obviously statements of differing cognitive value; $a = a$ holds a priori and, according to Kant, is to be labelled analytic, while statements of the form $a = b$ often contain very valuable extensions of our knowledge and cannot always be established a priori. The discovery that the rising sun is not new every morning, but always the same, was one of the most fertile astronomical discoveries. Even to-day the identification of a small planet or a comet is not always a matter of course. Now if we were to regard equality as a relation between that which the names 'a' and 'b' designate, it would seem that $a = b$ could not differ from $a = a$ (i.e. provided $a = b$ is true). A relation would thereby be expressed of a thing to itself, and indeed one in which each thing stands to itself but to no other thing. What is intended to be said by $a = b$ seems to be that the signs or names 'a' and 'b' designate the same thing, so that those signs themselves would be under discussion; a relation between them would be asserted. But this relation would hold between the names or signs only in so far as they named or designated something. It would be mediated by the connexion of each of the two signs with the same designated thing. But this is arbitrary. Nobody can be forbidden to use any arbitrarily producible event or object as a sign for something. In that case the sentence $a = b$ would no longer refer to the subject matter, but only to its mode of designation; we would express no proper knowledge by its means. But in many cases this is just what we want to do. If the sign 'a' is distinguished

From *Translations from the Philosophical Writings of Gottlob Frege*, ed. Peter Geach and Max Black (this piece translated by Max Black) (Oxford: Blackwell, 1952), 56–78. Reprinted by permission of Blackwell Publishers.

First published in *Zeitschrift für Philosophie und philosophische Kritik*, 100 (1892), 25–50.

[1] I use this word strictly and understand '$a = b$' to have the sense of 'a is the same as b' or 'a and b coincide'.

[2] Trans. note: The reference is to Frege's *Begriffsschrift, eine der arithmetischen nachgebildete Formelsprache des reinen Denkens* (Halle, 1879).

from the sign 'b' only as object (here, by means of its shape), not as sign (i.e. not by the manner in which it designates something), the cognitive value of $a = a$ becomes essentially equal to that of $a = b$, provided $a = b$ is true. A difference can arise only if the difference between the signs corresponds to a difference in the mode of presentation of that which is designated. Let a, b, c be the lines connecting the vertices of a triangle with the midpoints of the opposite sides. The point of intersection of a and b is then the same as the point of intersection of b and c. So we have different designations for the same point, and these names ('point of intersection of a and b,' 'point of intersection of b and c') likewise indicate the mode of presentation; and hence the statement contains actual knowledge.

It is natural, now, to think of there being connected with a sign (name, combination of words, letter), besides that to which the sign refers, which may be called the reference of the sign, also what I should like to call the *sense* of the sign, wherein the mode of presentation is contained. In our example, accordingly, the reference of the expressions 'the point of intersection of a and b' and 'the point of intersection of b and c' would be the same, but not their senses. The reference of 'evening star' would be the same as that of 'morning star,' but not the sense.

It is clear from the context that by 'sign' and 'name' I have here understood any designation representing a proper name, which thus has as its reference a definite object (this word taken in the widest range), but not a concept or a relation, which shall be discussed further in another article.[3] The designation of a single object can also consist of several words or other signs. For brevity, let every such designation be called a proper name.

The sense of a proper name is grasped by everybody who is sufficiently familiar with the language or totality of designations to which it belongs;[4] but this serves to illuminate only a single aspect of the reference, supposing it to have one. Comprehensive knowledge of the reference

[3] Trans. note: See his 'Ueber Begriff und Gegenstand', *Vierteljahrsschrift für wissenschaftliche Philosophie*, 16 (1892), 192–205.

[4] In the case of an actual proper name such as 'Aristotle' opinions as to the sense may differ. It might, for instance, be taken to be the following: the pupil of Plato and teacher of Alexander the Great. Anybody who does this will attach another sense to the sentence 'Aristotle was born in Stagira' than will a man who takes as the sense of the name: the teacher of Alexander the Great who was born in Stagira. So long as the reference remains the same, such variations of sense may be tolerated, although they are to be avoided in the theoretical structure of a demonstrative science and ought not to occur in a perfect language.

would require us to be able to say immediately whether any given sense belongs to it. To such knowledge we never attain.

The regular connexion between a sign, its sense, and its reference is of such a kind that to the sign there corresponds a definite sense and to that in turn a definite reference, while to a given reference (an object) there does not belong only a single sign. The same sense has different expressions in different languages or even in the same language. To be sure, exceptions to this regular behaviour occur. To every expression belonging to a complete totality of signs, there should certainly correspond a definite sense; but natural languages often do not satisfy this condition, and one must be content if the same word has the same sense in the same context. It may perhaps be granted that every grammatically well-formed expression representing a proper name always has a sense. But this is not to say that to the sense there also corresponds a reference. The words 'the celestial body most distant from the Earth' have a sense, but it is very doubtful if they also have a reference. The expression 'the least rapidly convergent series' has a sense; but it is known to have no reference, since for every given convergent series, another convergent, but less rapidly convergent, series can be found. In grasping a sense, one is not certainly assured of a reference.

If words are used in the ordinary way, what one intends to speak of is their reference. It can also happen, however, that one wishes to talk about the words themselves or their sense. This happens, for instance, when the words of another are quoted. One's own words then first designate words of the other speaker, and only the latter have their usual reference. We then have signs of signs. In writing, the words are in this case enclosed in quotation marks. Accordingly, a word standing between quotation marks must not be taken as having its ordinary reference.

In order to speak of the sense of an expression 'A' one may simply use the phrase 'the sense of the expression "A"'. In reported speech one talks about the sense, e.g., of another person's remarks. It is quite clear that in this way of speaking words do not have their customary reference but designate what is usually their sense. In order to have a short expression, we will say: In reported speech, words are used *indirectly* or have their *indirect* reference. We distinguish accordingly the *customary* from the *indirect* reference of a word; and its *customary* sense from its *indirect* sense. The indirect reference of a word is accordingly its customary sense. Such exceptions must always be borne in mind if the mode of connexion between sign, sense, and reference in particular cases is to be correctly understood.

The reference and sense of a sign are to be distinguished from the associated idea. If the reference of a sign is an object perceivable by the senses, my idea of it is an internal image,[5] arising from memories of sense impressions which I have had and acts, both internal and external, which I have performed. Such an idea is often saturated with feeling; the clarity of its separate parts varies and oscillates. The same sense is not always connected, even in the same man, with the same idea. The idea is subjective: one man's idea is not that of another. There result, as a matter of course, a variety of differences in the ideas associated with the same sense. A painter, a horseman, and a zoologist will probably connect different ideas with the name 'Bucephalus'. This constitutes an essential distinction between the idea and the sign's sense, which may be the common property of many and therefore is not a part of a mode of the individual mind. For one can hardly deny that mankind has a common store of thoughts which is transmitted from one generation to another.[6]

In the light of this, one need have no scruples in speaking simply of *the* sense, whereas in the case of an idea one must, strictly speaking, add to whom it belongs and at what time. It might perhaps be said: Just as one man connects this idea, and another that idea, with the same word, so also one man can associate this sense and another that sense. But there still remains a difference in the mode of connexion. They are not prevented from grasping the same sense; but they cannot have the same idea. *Si duo idem faciunt, non est idem.* If two persons picture the same thing, each still has his own idea. It is indeed sometimes possible to establish differences in the ideas, or even in the sensations, of different men; but an exact comparison is not possible, because we cannot have both ideas together in the same consciousness.

The reference of a proper name is the object itself which we designate by its means; the idea, which we have in that case, is wholly subjective; in between lies the sense, which is indeed no longer subjective like the idea, but is yet not the object itself. The following analogy will perhaps clarify these relationships. Somebody observes the Moon through a telescope. I compare the Moon itself to the reference; it is the object of the observation, mediated by the real image projected by the object

[5] We can include with ideas the direct experiences in which sense-impressions and acts themselves take the place of the traces which they have left in the mind. The distinction is unimportant for our purpose, especially since memories of sense-impressions and acts always help to complete the perceptual image. One can also understand direct experience as including any object, in so far as it is sensibly perceptible or spatial.

[6] Hence it is inadvisable to use the word 'idea' to designate something so basically different.

glass in the interior of the telescope, and by the retinal image of the observer. The former I compare to the sense, the latter is like the idea or experience. The optical image in the telescope is indeed one-sided and dependent upon the standpoint of observation; but it is still objective, inasmuch as it can be used by several observers. At any rate it could be arranged for several to use it simultaneously. But each one would have his own retinal image. On account of the diverse shapes of the observers' eyes, even a geometrical congruence could hardly be achieved, and an actual coincidence would be out of the question. This analogy might be developed still further, by assuming A's retinal image made visible to B; or A might also see his own retinal image in a mirror. In this way we might perhaps show how an idea can itself be taken as an object, but as such is not for the observer what it directly is for the person having the idea. But to pursue this would take us too far afield.

We can now recognize three levels of difference between words, expressions, or whole sentences. The difference may concern at most the ideas, or the sense but not the reference, or, finally, the reference as well. With respect to the first level, it is to be noted that, on account of the uncertain connexion of ideas with words, a difference may hold for one person, which another does not find. The difference between a translation and the original text should properly not overstep the first level. To the possible differences here belong also the colouring and shading which poetic eloquence seeks to give to the sense. Such colouring and shading are not objective, and must be evoked by each hearer or reader according to the hints of the poet or the speaker. Without some affinity in human ideas art would certainly be impossible; but it can never be exactly determined how far the intentions of the poet are realized.

In what follows there will be no further discussion of ideas and experiences; they have been mentioned here only to ensure that the idea aroused in the hearer by a word shall not be confused with its sense or its reference.

To make short and exact expressions possible, let the following phraseology be established:

A proper name (word, sign, sign combination, expression) *expresses* its sense, *stands for* or *designates* its reference. By means of a sign we express its sense and designate its reference.

Idealists or sceptics will perhaps long since have objected: 'You talk, without further ado, of the Moon as an object; but how do you know that the name 'the Moon' has any reference? How do you know that anything whatsoever has a reference?' I reply that when we say 'the

Moon,' we do not intend to speak of our idea of the Moon, nor are we satisfied with the sense alone, but we presuppose a reference. To assume that in the sentence 'The Moon is smaller than the Earth' the idea of the Moon is in question, would be flatly to misunderstand the sense. If this is what the speaker wanted, he would use the phrase 'my idea of the Moon'. Now we can of course be mistaken in the presupposition, and such mistakes have indeed occurred. But the question whether the presupposition is perhaps always mistaken need not be answered here; in order to justify mention of the reference of a sign it is enough, at first, to point out our intention in speaking or thinking. (We must then add the reservation: provided such reference exists.)

So far we have considered the sense and reference only of such expressions, words, or signs as we have called proper names. We now inquire concerning the sense and reference for an entire declarative sentence. Such a sentence contains a thought.[7] Is this thought, now, to be regarded as its sense or its reference? Let us assume for the time being that the sentence has reference. If we now replace one word of the sentence by another having the same reference, but a different sense, this can have no bearing upon the reference of the sentence. Yet we can see that in such a case the thought changes; since, e.g., the thought in the sentence 'The morning star is a body illuminated by the Sun' differs from that in the sentence 'The evening star is a body illuminated by the Sun.' Anybody who did not know that the evening star is the morning star might hold the one thought to be true, the other false. The thought, accordingly, cannot be the reference of the sentence, but must rather be considered as the sense. What is the position now with regard to the reference? Have we a right even to inquire about it? Is it possible that a sentence as a whole has only a sense, but no reference? At any rate, one might expect that such sentences occur, just as there are parts of sentences having sense but no reference. And sentences which contain proper names without reference will be of this kind. The sentence 'Odysseus was set ashore at Ithaca while sound asleep' obviously has a sense. But since it is doubtful whether the name 'Odysseus', occurring therein, has reference, it is also doubtful whether the whole sentence has one. Yet it is certain, nevertheless, that anyone who seriously took the sentence to be true or false would ascribe to the name 'Odysseus' a reference, not merely a sense; for it is of the reference of the name that the predicate is affirmed or denied. Whoever does not admit the name

[7] By a thought I understand not the subjective performance of thinking but its objective content, which is capable of being the common property of several thinkers.

has reference can neither apply nor withhold the predicate. But in that case it would be superfluous to advance to the reference of the name; one could be satisfied with the sense, if one wanted to go no further than the thought. If it were a question only of the sense of the sentence, the thought, it would be unnecessary to bother with the reference of a part of the sentence; only the sense, not the reference, of the part is relevant to the sense of the whole sentence. The thought remains the same whether 'Odysseus' has reference or not. The fact that we concern ourselves at all about the reference of a part of the sentence indicates that we generally recognize and expect a reference for the sentence itself. The thought loses value for us as soon as we recognize that the reference of one of its parts is missing. We are therefore justified in not being satisfied with the sense of a sentence, and in inquiring also as to its reference. But now why do we want every proper name to have not only a sense, but also a reference? Why is the thought not enough for us? Because, and to the extent that, we are concerned with its truth value. This is not always the case. In hearing an epic poem, for instance, apart from the euphony of the language we are interested only in the sense of the sentences and the images and feelings thereby aroused. The question of truth would cause us to abandon aesthetic delight for an attitude of scientific investigation. Hence it is a matter of no concern to us whether the name 'Odysseus', for instance, has reference, so long as we accept the poem as a work of art.[8] It is the striving for truth that drives us always to advance from the sense to the reference.

We have seen that the reference of a sentence may always be sought, whenever the reference of its components is involved; and that this is the case when and only when we are inquiring after the truth value.

We are therefore driven into accepting the *truth value* of a sentence as constituting its reference. By the truth value of a sentence I understand the circumstance that it is true or false. There are no further truth values. For brevity I call the one the True, the other the False. Every declarative sentence concerned with the reference of its words is therefore to be regarded as a proper name, and its reference, if it has one, is either the True or the False. These two objects are recognized, if only implicitly, by everybody who judges something to be true—and so even by a sceptic. The designation of the truth values as objects may appear to be an arbitrary fancy or perhaps a mere play upon words, from which no profound consequences could be drawn. What I mean by an

[8] It would be desirable to have a special term for signs having only sense. If we name them, say, representations, the words of the actors on the stage would be representations; indeed the actor himself would be a representation.

object can be more exactly discussed only in connexion with concept and relation. I will reserve this for another article.[9] But so much should already be clear, that in every judgment,[10] no matter how trivial, the step from the level of thoughts to the level of reference (the objective) has already been taken.

One might be tempted to regard the relation of the thought to the True not as that of sense to reference, but rather as that of subject to predicate. One can, indeed, say: 'The thought, that 5 is a prime number, is true.' But closer examination shows that nothing more has been said than in the simple sentence '5 is a prime number.' The truth claim arises in each case from the form of the declarative sentence, and when the latter lacks its usual force, e.g., in the mouth of an actor upon the stage, even the sentence 'The thought that 5 is a prime number is true' contains only a thought, and indeed the same thought as the simple '5 is a prime number.' It follows that the relation of the thought to the True may not be compared with that of subject to predicate. Subject and predicate (understood in the logical sense) are indeed elements of thought; they stand on the same level for knowledge. By combining subject and predicate, one reaches only a thought, never passes from sense to reference, never from a thought to its truth value. One moves at the same level but never advances from one level to the next. A truth value cannot be a part of a thought, any more than, say, the Sun can, for it is not a sense but an object.

If our supposition that the reference of a sentence is its truth value is correct, the latter must remain unchanged when a part of the sentence is replaced by an expression having the same reference. And this is in fact the case. Leibniz gives the definition: '*Eadem sunt, quae sibi mutuo substitui possunt, salva veritate*.' What else but the truth value could be found, that belongs quite generally to every sentence if the reference of its components is relevant, and remains unchanged by substitutions of the kind in question?

If now the truth value of a sentence is its reference, then on the one hand all true sentences have the same reference and so, on the other hand, do all false sentences. From this we see that in the reference of the sentence all that is specific is obliterated. We can never be concerned only with the reference of a sentence; but again the mere thought alone

[9] Trans. note: See his 'Ueber Begriff und Gegenstand', *Vierteljahrsschrift für wissenschaftliche Philosophie*, 16 (1892), 192–205.

[10] A judgment, for me is not the mere comprehension of a thought, but the admission of its truth.

yields no knowledge, but only the thought together with its reference, i.e. its truth value. Judgments can be regarded as advances from a thought to a truth value. Naturally this cannot be a definition. Judgment is something quite peculiar and incomparable. One might also say that judgments are distinctions of parts within truth values. Such distinction occurs by a return to the thought. To every sense belonging to a truth value there would correspond its own manner of analysis. However, I have here used the word 'part' in a special sense. I have in fact transferred the relation between the parts and the whole of the sentence to its reference, by calling the reference of a word part of the reference of the sentence, if the word itself is a part of the sentence. This way of speaking can certainly be attacked, because the whole reference and one part of it do not suffice to determine the remainder, and because the word 'part' is already used in another sense of bodies. A special term would need to be invented.

The supposition that the truth value of a sentence is its reference shall now be put to further test. We have found that the truth value of a sentence remains unchanged when an expression is replaced by another having the same reference: but we have not yet considered the case in which the expression to be replaced is itself a sentence. Now if our view is correct, the truth value of a sentence containing another as part must remain unchanged when the part is replaced by another sentence having the same truth value. Exceptions are to be expected when the whole sentence or its part is direct or indirect quotation; for in such cases, as we have seen, the words do not have their customary reference. In direct quotation, a sentence designates another sentence, and in indirect quotation a thought.

We are thus led to consider subordinate sentences or clauses. These occur as parts of a sentence complex, which is, from the logical standpoint, likewise a sentence—a main sentence. But here we meet the question whether it is also true of the subordinate sentence that its reference is a truth value. Of indirect quotation we already know the opposite. Grammarians view subordinate clauses as representatives of parts of sentences and divide them accordingly into noun clauses, adjective clauses, adverbial clauses. This might generate the supposition that the reference of a subordinate clause was not a truth value but rather of the same kind as the reference of a noun or adjective or adverb—in short, of a part of a sentence, whose sense was not a thought but only a part of a thought. Only a more thorough investigation can clarify the issue. In so doing, we shall not follow the grammatical categories strictly, but rather group together what is logically of the same kind. Let us first search for cases in which the sense of the subordinate clause, as we have just supposed, is not an independent thought.

The case of an abstract[11] noun clause, introduced by 'that', includes the case of indirect quotation, in which we have seen the words to have their indirect reference coinciding with what is customarily their sense. In this case, then, the subordinate clause has for its reference a thought, not a truth value; as sense not a thought, but the sense of the words 'the thought, that . . . ,' which is only a part of the thought in the entire complex sentence. This happens after 'say', 'hear', 'be of the opinion', 'be convinced', 'conclude', and similar words.[12] There is a different, and indeed somewhat complicated, situation after words like 'perceive', 'know', 'fancy', which are to be considered later.

That in the cases of the first kind the reference of the subordinate clause is in fact the thought can also be recognized by seeing that it is indifferent to the truth of the whole whether the subordinate clause is true or false. Let us compare, for instance, the two sentences 'Copernicus believed that the planetary orbits are circles' and 'Copernicus believed that the apparent motion of the Sun is produced by the real motion of the Earth.' One subordinate clause can be substituted for the other without harm to the truth. The main clause and the subordinate clause together have as their sense only a single thought, and the truth of the whole includes neither the truth nor the untruth of the subordinate clause. In such cases it is not permissible to replace one expression in the subordinate clause by another having the same customary reference, but only by one having the same indirect reference, i.e. the same customary sense. If somebody were to conclude: The reference of a sentence is not its truth value, for in that case it could always be replaced by another sentence of the same truth value; he would prove too much; one might just as well claim that the reference of 'morning star' is not Venus, since one may not always say 'Venus' in place of 'morning star'. One has the right to conclude only that the reference of a sentence is not *always* its truth value, and that 'morning star' does not always stand for the planet Venus, viz. when the word has its indirect reference. An exception of such a kind occurs in the subordinate clause just considered which has a thought as its reference.

If one says 'It seems that . . .' one means 'It seems to me that . . .' or 'I think that . . .' We therefore have the same case again. The situation is

[11] Trans. note: A literal translation of Frege's 'abstracten Nennsätzen' whose meaning eludes me.

[12] In 'A lied in saying he had seen B', the subordinate clause designates a thought which is said (1) to have been asserted by A (2) while A was convinced of its falsity.

similar in the case of expressions such as 'to be pleased', 'to regret', 'to approve', 'to blame', 'to hope', 'to fear'. If, toward the end of the battle of Waterloo,[13] Wellington was glad that the Prussians were coming, the basis for his joy was a conviction. Had he been deceived, he would have been no less pleased so long as his illusion lasted; and before he became so convinced he could not have been pleased that the Prussians were coming—even though in fact they might have been already approaching.

Just as a conviction or a belief is the ground of a feeling, it can, as in inference, also be the ground of a conviction. In the sentence: 'Columbus inferred from the roundness of the Earth that he could reach India by travelling towards the west,' we have as the reference of the parts two thoughts, that the Earth is round, and that Columbus by travelling to the west could reach India. All that is relevant here is that Columbus was convinced of both, and that the one conviction was a ground for the other. Whether the Earth is really round, and whether Columbus could really reach India by travelling to the west, are immaterial to the truth of our sentence; but it is not immaterial whether we replace 'the Earth' by 'the planet which is accompanied by a moon whose diameter is greater than the fourth part of its own.' Here also we have the indirect reference of the words.

Adverbial final clauses beginning 'in order that' also belong here; for obviously the purpose is a thought; therefore: indirect reference for the words, subjunctive mood.

A subordinate clause with 'that' after 'command', 'ask', 'forbid', would appear in direct speech as an imperative. Such a clause has no reference but only a sense. A command, a request, are indeed not thoughts, yet they stand on the same level as thoughts. Hence in subordinate clauses depending upon 'command', 'ask', etc., words have their indirect reference. The reference of such a clause is therefore not a truth value but a command, a request, and so forth.

The case is similar for the dependent question in phrases such as 'doubt whether', 'not to know what'. It is easy to see that here also the words are to be taken to have their indirect reference. Dependent clauses expressing questions and beginning with 'who', 'what', 'where', 'when', 'how', 'by what means', etc., seem at times to approximate very closely to adverbial clauses in which words have their customary references. These cases are distinguished linguistically [in German] by the mood of the verb. With the subjunctive, we have a dependent

[13] Trans. note: Frege uses the Prussian name for the battle—'Belle Alliance'.

question and indirect reference of the words, so that a proper name cannot in general be replaced by another name of the same object.

In the cases so far considered the words of the subordinate clauses had their indirect reference, and this made it clear that the reference of the subordinate clause itself was indirect, i.e. not a truth value but a thought, a command, a request, a question. The subordinate clause could be regarded as a noun, indeed one could say: as a proper name of that thought, that command, etc., which it represented in the context of the sentence structure.

We now come to other subordinate clauses, in which the words do have their customary reference without however a thought occurring as sense and a truth value as reference. How this is possible is best made clear by examples.

Whoever discovered the elliptic form of the planetary orbits died in misery.

If the sense of the subordinate clause were here a thought, it would have to be possible to express it also in a separate sentence. But this does not work, because the grammatical subject 'whoever' has no independent sense and only mediates the relation with the consequent clause 'died in misery'. For this reason the sense of the subordinate clause is not a complete thought, and its reference is Kepler, not a truth value. One might object that the sense of the whole does contain a thought as part, viz. that there was somebody who first discovered the elliptic form of the planetary orbits; for whoever takes the whole to be true cannot deny this part. This is undoubtedly so; but only because otherwise the dependent clause 'whoever discovered the elliptic form of the planetary orbits' would have no reference. If anything is asserted there is always an obvious presupposition that the simple or compound proper names used have reference. If one therefore asserts 'Kepler died in misery,' there is a presupposition that the name 'Kepler' designates something; but it does not follow that the sense of the sentence 'Kepler died in misery' contains the thought that the name 'Kepler' designates something. If this were the case the negation would have to run not

Kepler did not die in misery

but

Kepler did not die in misery, or the name 'Kepler' has no reference.

That the name 'Kepler' designates something is just as much a presupposition for the assertion

Kepler died in misery

as for the contrary assertion. Now languages have the fault of containing expressions which fail to designate an object (although their grammatical form seems to qualify them for that purpose) because the truth of some sentences is a prerequisite. Thus it depends on the truth of the sentence:

> There was someone who discovered the elliptic form of the planetary orbits

whether the subordinate clause

> Whoever discovered the elliptic form of the planetary orbits

really designates an object or only seems to do so while having in fact no reference. And thus it may appear as if our subordinate clause contained as a part of its sense the thought that there was somebody who discovered the elliptic form of the planetary orbits. If this were right the negation would run:

> Either whoever discovered the elliptic form of the planetary orbits did not die in misery or there was nobody who discovered the elliptic form of the planetary orbits.

This arises from an imperfection of language, from which even the symbolic language of mathematical analysis is not altogether free; even there combinations of symbols can occur that seem to stand for something but have (at least so far) no reference, e.g. divergent infinite series. This can be avoided, e.g., by means of the special stipulation that divergent infinite series shall stand for the number 0. A logically perfect language (*Begriffsschrift*) should satisfy the conditions, that every expression grammatically well constructed as a proper name out of signs already introduced shall in fact designate an object, and that no new sign shall be introduced as a proper name without being secured a reference. The logic books contain warnings against logical mistakes arising from the ambiguity of expressions. I regard as no less pertinent a warning against apparent proper names having no reference. The history of mathematics supplies errors which have arisen in this way. This lends itself to demagogic abuse as easily as ambiguity—perhaps more easily. 'The will of the people' can serve as an example; for it is easy to establish that there is at any rate no generally accepted reference for this expression. It is therefore by no means unimportant to eliminate the source of these mistakes, at least in science, once and for all. Then such objections as the one discussed above would become impossible, because it could never depend upon the truth of a thought whether a proper name had a reference.

With the consideration of these noun clauses may be coupled that of types of adjective and adverbial clauses which are logically in close relation to them.

Adjective clauses also serve to construct compound proper names, though, unlike noun clauses, they are not sufficient by themselves for this purpose. These adjective clauses are to be regarded as equivalent to adjectives. Instead of 'the square root of 4 which is smaller than 0', one can also say 'the negative square root of 4'. We have here the case of a compound proper name constructed from the expression for a concept with the help of the singular definite article. This is at any rate permissible if the concept applies to one and only one single object.[14]

Expressions for concepts can be so constructed that marks of a concept are given by adjective clauses as, in our example, by the clause 'which is smaller than 0'. It is evident that such an adjective clause cannot have a thought as sense or a truth value as reference, any more than the noun clause could. Its sense, which can also be expressed in many cases by a single adjective, is only a part of a thought. Here, as in the case of the noun clause, there is no independent subject and therefore no possibility of reproducing the sense of the subordinate clause in an independent sentence.

Places, instants, stretches of time, are, logically considered, objects; hence the linguistic designation of a definite place, a definite instant, or a stretch of time is to be regarded as a proper name. Now adverbial clauses of place and time can be used for the construction of such a proper name in a manner similar to that which we have seen in the case of noun and adjective clauses. In the same way, expressions for concepts bringing in places, etc., can be constructed. It is to be noted here also that the sense of these subordinate clauses cannot be reproduced in an independent sentence, since an essential component, viz. the determination of place or time, is missing and is only indicated by a relative pronoun or a conjunction.[15]

[14] In accordance with what was said above, an expression of the kind in question must actually always be assured of reference, by means of a special stipulation, e.g. by the convention that 0 shall count as its reference, when the concept applies to no object or to more than one.

[15] In the case of these sentences, various interpretations are easily possible. The sense of the sentence, 'After Schleswig-Holstein was separated from Denmark, Prussia and Austria quarrelled' can also be rendered in the form 'After the separation of Schleswig-Holstein from Denmark, Prussia and Austria quarrelled.' In this version, it is surely sufficiently clear that the sense is not to be taken as having as a part the thought that Schleswig-Holstein was once separated from Denmark, but that this is the necessary presupposition in order for the expression 'after the separation of Schleswig-Holstein from Denmark' to have any reference at all. To be sure, our sentence can also be interpreted as saying that Schleswig-Holstein was once separated from Denmark. We then have a case which is to be considered later. In order to understand the difference more clearly, let us project ourselves into the mind of a Chinese

In conditional clauses, also, there may usually be recognized to occur an indefinite indicator, having a similar correlate in the dependent clause. (We have already seen this occur in noun, adjective, and adverbial clauses.) In so far as each indicator refers to the other, both clauses together form a connected whole, which as a rule expresses only a single thought. In the sentence

> If a number is less than 1 and greater than 0, its square is less than 1 and greater than 0

the component in question is 'a number' in the conditional clause and 'its' in the dependent clause. It is by means of this very indefiniteness that the sense acquires the generality expected of a law. It is this which is responsible for the fact that the antecedent clause alone has no complete thought as its sense and in combination with the consequent clause expresses one and only one thought, whose parts are no longer thoughts. It is, in general, incorrect to say that in the hypothetical judgment two judgments are put in reciprocal relationship. If this or something similar is said, the word 'judgment' is used in the same sense as I have connected with the word 'thought', so that I would use the formulation: 'A hypothetical thought establishes a reciprocal relationship between two thoughts.' This could be true only if an indefinite indicator is absent;[16] but in such a case there would also be no generality.

If an instant of time is to be indefinitely indicated in both conditional and dependent clauses, this is often achieved merely by using the present tense of the verb, which in such a case however does not indicate the temporal present. This grammatical form is then the indefinite indicator in the main and subordinate clauses. An example of this is: 'When the Sun is in the tropic of Cancer, the longest day in the northern hemisphere occurs.' Here, also, it is impossible to express the sense of the subordinate clause in a full sentence, because this sense is not a complete thought. If we say: 'The Sun is in the tropic of Cancer,' this would refer to our present time and thereby change the sense. Just as little is the sense of the main clause a thought; only the whole, composed of main and subordinate clauses, has such a sense. It may be added that

who, having little knowledge of European history, believes it to be false that Schleswig-Holstein was ever separated from Denmark. He will take our sentence, in the first version, to be neither true nor false but will deny it to have any reference, on the ground of absence of reference for its subordinate clause. This clause would only apparently determine a time. If he interpreted our sentence in the second way, however, he would find a thought expressed in it which he would take to be false, beside a part which would be without reference for him.

[16] At times an explicit linguistic indication is missing and must be read off from the entire context.

several common components in the antecedent and consequent clauses may be indefinitely indicated.

It is clear that noun clauses with 'who' or 'what' and adverbial clauses with 'where', 'when', 'wherever', 'whenever' are often to be interpreted as having the sense of conditional clauses, e.g. 'who touches pitch, defiles himself.'

Adjective clauses can also take the place of conditional clauses. Thus the sense of the sentence previously used can be given in the form 'The square of a number which is less than 1 and greater than 0 is less than 1 and greater than 0.'

The situation is quite different if the common component of the two clauses is designated by a proper name. In the sentence:

> Napoleon, who recognized the danger to his right flank, himself led his guards against the enemy position

two thoughts are expressed:

1. Napoleon recognized the danger to his right flank
2. Napoleon himself led his guards against the enemy position.

When and where this happened is to be fixed only by the context, but is nevertheless to be taken as definitely determined thereby. If the entire sentence is uttered as an assertion, we thereby simultaneously assert both component sentences. If one of the parts is false, the whole is false. Here we have the case that the subordinate clause by itself has a complete thought as sense (if we complete it by indication of place and time). The reference of the subordinate clause is accordingly a truth value. We can therefore expect that it may be replaced, without harm to the truth value of the whole, by a sentence having the same truth value. This is indeed the case; but it is to be noticed that for purely grammatical reasons, its subject must be 'Napoleon', for only then can it be brought into the form of an adjective clause belonging to 'Napoleon'. But if the demand that it be expressed in this form be waived, and the connexion be shown by 'and', this restriction disappears.

Subsidiary clauses beginning with 'although' also express complete thoughts. This conjunction actually has no sense and does not change the sense of the clause but only illuminates it in a peculiar fashion.[17] We could indeed replace the conditional clause without harm to the truth of the whole by another of the same truth value; but the light in which the clause is placed by the conjunction might then easily appear unsuitable, as if a song with a sad subject were to be sung in a lively fashion.

[17] Similarly in the case of 'but', 'yet'.

In the last cases the truth of the whole included the truth of the component clauses. The case is different if a conditional clause expresses a complete thought by containing, in place of an indefinite indicator, a proper name or something which is to be regarded as equivalent. In the sentence

If the Sun has already risen, the sky is very cloudy

the time is the present, that is to say, definite. And the place is also to be thought of as definite. Here it can be said that a relation between the truth values of conditional and dependent clauses has been asserted, viz. such that the case does not occur in which the antecedent stands for the True and the consequent for the False. Accordingly, our sentence is true if the Sun has not yet risen, whether the sky is very cloudy or not, and also if the Sun has risen and the sky is very cloudy. Since only truth values are here in question, each component clause can be replaced by another of the same truth value without changing the truth value of the whole. To be sure, the light in which the subject then appears would usually be unsuitable; the thought might easily seem distorted; but this has nothing to do with its truth value. One must always take care not to clash with the subsidiary thoughts, which are however not explicitly expressed and therefore should not be reckoned in the sense. Hence, also, no account need be taken of their truth values.[18]

The simple cases have now been discussed. Let us review what we have learned.

The subordinate clause usually has for its sense not a thought, but only a part of one, and consequently no truth value as reference. The reason for this is either that the words in the subordinate clause have indirect reference, so that the reference, not the sense, of the subordinate clause is a thought; or else that, on account of the presence of an indefinite indicator, the subordinate clause is incomplete and expresses a thought only when combined with the main clause. It may happen, however, that the sense of the subsidiary clause is a complete thought, in which case it can be replaced by another of the same truth value without harm to the truth of the whole—provided there are no grammatical obstacles.

An examination of all the subordinate clauses which one may encounter will soon provide some which do not fit well into these categories. The reason, so far as I can see, is that these subordinate clauses have no such simple sense. Almost always, it seems, we connect

[18] The thought of our sentence might also be expressed thus: 'Either the Sun has not risen yet or the sky is very cloudy'—which shows how this kind of sentence connexion is to be understood.

with the main thoughts expressed by us subsidiary thoughts which, although not expressed, are associated with our words, in accordance with psychological laws, by the hearer. And since the subsidiary thought appears to be connected with our words of its own accord, almost like the main thought itself, we want it also to be expressed. The sense of the sentence is thereby enriched, and it may well happen that we have more simple thoughts than clauses. In many cases the sentence must be understood in this way, in others it may be doubtful whether the subsidiary thought belongs to the sense of the sentence or only accompanies it.[19] One might perhaps find that the sentence

> Napoleon, who recognized the danger to his right flank, himself led his guards against the enemy position

expresses not only the two thoughts shown above, but also the thought that the knowledge of the danger was the reason why he led the guards against the enemy position. One may in fact doubt whether this thought is merely slightly suggested or really expressed. Let the question be considered whether our sentence be false if Napoleon's decision had already been made before he recognized the danger. If our sentence could be true in spite of this, the subsidiary thought should not be understood as part of the sense. One would probably decide in favour of this. The alternative would make for a quite complicated situation: We would have more simple thoughts than clauses. If the sentence

> Napoleon recognized the danger to his right flank

were now to be replaced by another having the same truth value, e.g.

> Napoleon was already more than 45 years old

not only would our first thought be changed, but also our third one. Hence the truth value of the latter might change—viz. if his age was not the reason for the decision to lead the guards against the enemy. This shows why clauses of equal truth value cannot always be substituted for one another in such cases. The clause expresses more through its connexion with another than it does in isolation.

Let us now consider cases where this regularly happens. In the sentence:

> Bebel mistakenly supposes that the return of Alsace-Lorraine would appease France's desire for revenge

two thoughts are expressed, which are not however shown by means of antecedent and consequent clauses, viz.:

[19] This may be important for the question whether an assertion is a lie, or an oath a perjury.

(1) Bebel believes that the return of Alsace-Lorraine would appease France's desire for revenge
(2) the return of Alsace-Lorraine would not appease France's desire for revenge.

In the expression of the first thought, the words of the subordinate clause have their indirect reference, while the same words have their customary reference in the expression of the second thought. This shows that the subordinate clause in our original complex sentence is to be taken twice over, with different reference, standing once for a thought, once for a truth value. Since the truth value is not the whole reference of the subordinate clause, we cannot simply replace the latter by another of equal truth value. Similar considerations apply to expressions such as 'know', 'discover', 'it is known that'.

By means of a subordinate causal clause and the associated main clause we express several thoughts, which however do not correspond separately to the original clauses. In the sentence: 'Because ice is less dense than water, it floats on water' we have

(1) Ice is less dense than water;
(2) If anything is less dense than water, it floats on water;
(3) Ice floats on water.

The third thought, however, need not be explicitly introduced, since it is contained in the remaining two. On the other hand, neither the first and third nor the second and third combined would furnish the sense of our sentence. It can now be seen that our subordinate clause

because ice is less dense than water

expresses our first thought, as well as a part of our second. This is how it comes to pass that our subsidiary clause cannot be simply replaced by another of equal truth value; for this would alter our second thought and thereby might well alter its truth value.

The situation is similar in the sentence

If iron were less dense than water, it would float on water.

Here we have the two thoughts that iron is not less dense than water, and that something floats on water if it is less dense than water. The subsidiary clause again expresses one thought and a part of the other.

If we interpret the sentence already considered

After Schleswig-Holstein was separated from Denmark, Prussia and Austria quarrelled

in such a way that it expresses the thought that Schleswig-Holstein was once separated from Denmark, we have first this thought, and secondly the thought that at a time, more closely determined by the subordinate clause, Prussia and Austria quarrelled. Here also the subordinate clause expresses not only one thought but also a part of another. Therefore it may not in general be replaced by another of the same truth value.

It is hard to exhaust all the possibilities given by language; but I hope to have brought to light at least the essential reasons why a subordinate clause may not always be replaced by another of equal truth value without harm to the truth of the whole sentence structure. These reasons arise:

(1) when the subordinate clause does not stand for a truth value, inasmuch as it expresses only a part of a thought;
(2) when the subordinate clause does stand for a truth value but is not restricted to so doing, inasmuch as its sense includes one thought and part of another.

The first case arises:

(a) in indirect reference of words
(b) if a part of the sentence is only an indefinite indicator instead of a proper name.

In the second case, the subsidiary clause may have to be taken twice over, viz. once in its customary reference, and the other time in indirect reference; or the sense of a part of the subordinate clause may likewise be a component of another thought, which, taken together with the thought directly expressed by the subordinate clause, makes up the sense of the whole sentence.

It follows with sufficient probability from the foregoing that the cases where a subordinate clause is not replaceable by another of the same value cannot be brought in disproof of our view that a truth value is the reference of a sentence having a thought as its sense.

Let us return to our starting point.

When we found '$a = a$' and '$a = b$' to have different cognitive values, the explanation is that for the purpose of knowledge, the sense of the sentence, viz., the thought expressed by it, is no less relevant than its reference, i.e. its truth value. If now $a = b$, then indeed the reference of 'b' is the same as that of 'a', and hence the truth value of '$a = b$' is the same as that of '$a = a$'. In spite of this, the sense of 'b' may differ from that of 'a', and thereby the sense expressed in '$a = b$' differs from that of '$a = a$'. In that case the two sentences do not have the same cognitive value. If we understand by 'judgment' the advance from the thought to its truth value, as in the above paper, we can also say that the judgments are different.

II

LETTER TO JOURDAIN

GOTTLOB FREGE

I do not believe that we can dispense with the sense of a name in logic; for a proposition must have a sense if it is to be useful. But a proposition consists of parts which must somehow contribute to the expression of the sense of the proposition; so they themselves must somehow have a sense. Take the proposition 'Etna is higher than Vesuvius'. This contains the name 'Etna', which occurs also in other propositions, e.g., in the proposition 'Etna is in Sicily'. The possibility of our understanding propositions which we have never heard before rests evidently on this, that we construct the sense of a proposition out of parts that correspond to the words. If we find the same word in two propositions, e.g., 'Etna', then we also recognize something common to the corresponding thoughts, something corresponding to this word. Without this, language in the proper sense would be impossible. We could indeed adopt the convention that certain signs were to express certain thoughts, like railway signals ('The track is clear'); but in this way we would always be restricted to a very narrow area, and we could not form a completely new proposition, one which would be understood by another person even though no special convention had been adopted beforehand for this case. Now that part of the thought which corresponds to the name 'Etna' cannot be Mount Etna itself; it cannot be the reference[1] of this name. For each individual piece of frozen, solidified lava which is part of Mount Etna would then also be part of the thought that Etna is higher than Vesuvius. But it seems to me absurd that pieces of lava, even pieces of which I had no knowledge, should be parts of my thought. Thus both things seem to me necessary: (1) the reference of a name, which is that

An extract from an undated letter, published in *Frege's Philosophical and Mathematical Correspondence*, ed. Gottfried Gabriel, Hans Hermes, Friedrich Kanbartel, Christian Thiel, and Albert Veraart, abridged for the English edn. by Brian McGuinness, and trans. Hans Kaal (Oxford: Blackwell, 1980). Reprinted by permission of Blackwell Publishers.

[1] Ed. note: Kaal's translation originally had 'meaning' here. 'Reference', however, brings the translation into line with the rest of this volume, including Frege's other essay, Essay I above. (The German word is '*Bedeutung*', which is usually rendered 'meaning', but which Frege is using in a technical way.) I have taken the liberty of replacing 'meaning', each time it occurs in Kaal's translation, by 'reference'.

about which something is being said, and (2) the sense of the name, which is part of the thought. Without reference, we could indeed have a thought, but only a mythological or literary thought, not a thought that could further scientific knowledge. Without a sense, we would have no thought, and hence also nothing that we could recognize as true.

To this can be added the following. Let us suppose an explorer travelling in an unexplored country sees a high snow-capped mountain on the northern horizon. By making inquiries among the natives he learns that its name is 'Aphla'. By sighting it from different points he determines its position as exactly as possible, enters it in a map, and writes in his diary: 'Aphla is at least 5000 metres high.' Another explorer sees a snow-capped mountain on the southern horizon and learns that it is called Ateb. He enters it in his map under this name. Later comparison shows that both explorers saw the same mountain. Now the content of the proposition 'Ateb is Aphla' is far from being a mere consequence of the principle of identity, but contains a valuable piece of geographical knowledge. What is stated in the proposition 'Ateb is Aphla' is certainly not the same thing as the content of the proposition 'Ateb is Ateb'. Now if what corresponded to the name 'Aphla' as part of the thought was the reference of the name and hence the mountain itself, then this would be the same in both thoughts. The thought expressed in the proposition 'Ateb is Aphla' would have to coincide with the one in 'Ateb is Ateb', which is far from being the case. What corresponds to the name 'Ateb' as part of the thought must therefore be different from what corresponds to the name 'Aphla' as part of the thought. This cannot therefore be the reference which is the same for both names, but must be something which is different in the two cases, and I say accordingly that the sense of the name 'Ateb' is different from the sense of the name 'Aphla'. Accordingly, the sense of the proposition 'Ateb is at least 5000 metres high' is also different from the sense of the proposition 'Aphla is at least 5000 metres high'. Someone who takes the latter to be true need not therefore take the former to be true. An object can be determined in different ways, and every one of these ways of determining it can give rise to a special name, and these different names then have different senses; for it is not self-evident that it is the same object which is being determined in different ways. We find this in astronomy in the case of planetoids and comets. Now if the sense of a name was something subjective, then the sense of the proposition in which the name occurs, and hence the thought, would also be something subjective, and the thought one man connects with this proposition would be different from the thought another man connects with it; a common store of thoughts, a common science would be impossible. It would be impossible for something one man said to contradict what another man said, because the two would not express the same thought at all, but each his own.

For these reasons I believe that the sense of a name is not something subjective [crossed out: in one's mental life], that it does not therefore belong to psychology, and that it is indispensable.

III

DESCRIPTIONS

BERTRAND RUSSELL

A 'description' may be of two sorts, definite and indefinite (or ambiguous). An indefinite description is a phrase of the form 'a so-and-so', and a definite description is a phrase of the form 'the so-and-so' (in the singular). Let us begin with the former.

'Who did you meet?' 'I met a man.' 'That is a very indefinite description.' We are therefore not departing from usage in our terminology. Our question is: What do I really assert when I assert 'I met a man'? Let us assume, for the moment, that my assertion is true, and that in fact I met Jones. It is clear that what I assert is *not* 'I met Jones.' I may say 'I met a man, but it was not Jones'; in that case, though I lie, I do not contradict myself, as I should do if when I say I met a man I really mean that I met Jones. It is clear also that the person to whom I am speaking can understand what I say, even if he is a foreigner and has never heard of Jones.

But we may go further: not only Jones, but no actual man, enters into my statement. This becomes obvious when the statement is false, since then there is no more reason why Jones should be supposed to enter into the proposition than why anyone else should. Indeed the statement would remain significant, though it could not possibly be true, even if there were no man at all. 'I met a unicorn' or 'I met a sea-serpent' is a perfectly significant assertion, if we know what it would be to be a unicorn or a sea-serpent, i.e. what is the definition of these fabulous monsters. Thus it is only what we may call the *concept* that enters into the proposition. In the case of 'unicorn', for example, there is only the concept: there is not also, somewhere among the shades, something unreal which may be called 'a unicorn'. Therefore, since it is significant (though false) to say 'I met a unicorn', it is clear that this proposition, rightly analysed, does not contain a constituent 'a unicorn', though it does contain the concept 'unicorn'.

The question of 'unreality', which confronts us at this point, is a very important one. Misled by grammar, the great majority of those logicians

An extract from Chapter XVI of Russell's *Introduction to Mathematical Philosophy* (London: Allen & Unwin, 1919). Reprinted by permission of Routledge.

who have dealt with this question have dealt with it on mistaken lines. They have regarded grammatical form as a surer guide in analysis than, in fact, it is. And they have not known what differences in grammatical form are important. 'I met Jones' and 'I met a man' would count traditionally as propositions of the same form, but in actual fact they are of quite different forms: the first names an actual person, Jones; while the second involves a propositional function, and becomes, when made explicit: 'The function "I met x and x is human" is sometimes true.' (It will be remembered that we adopted the convention of using 'sometimes' as not implying more than once.) This proposition is obviously not of the form 'I met x', which accounts for the existence of the proposition 'I met a unicorn' in spite of the fact that there is no such thing as 'a unicorn'.

For want of the apparatus of propositional functions, many logicians have been driven to the conclusion that there are unreal objects. It is argued, e.g. by Meinong,[1] that we can speak about 'the golden mountain', 'the round square', and so on; we can make true propositions of which these are the subjects; hence they must have some kind of logical being, since otherwise the propositions in which they occur would be meaningless. In such theories, it seems to me, there is a failure of that feeling for reality which ought to be preserved even in the most abstract studies. Logic, I should maintain, must no more admit a unicorn than zoology can; for logic is concerned with the real world just as truly as zoology, though with its more abstract and general features. To say that unicorns have an existence in heraldry, or in literature, or in imagination, is a most pitiful and paltry evasion. What exists in heraldry is not an animal, made of flesh and blood, moving and breathing of its own initiative. What exists is a picture, or a description in words. Similarly, to maintain that Hamlet, for example, exists in his own world, namely, in the world of Shakespeare's imagination, just as truly as (say) Napoleon existed in the ordinary world, is to say something deliberately confusing, or else confused to a degree which is scarcely credible. There is only one world, the 'real' world: Shakespeare's imagination is part of it, and the thoughts that he had in writing Hamlet are real. So are the thoughts that we have in reading the play. But it is of the very essence of fiction that only the thoughts, feelings, etc., in Shakespeare and his readers are real, and that there is not, in addition to them, an objective Hamlet. When you have taken account of all the feelings roused by Napoleon in writers and readers of history, you have not touched the actual man; but in the case of Hamlet you have come to the end of him.

[1] *Untersuchungen zur Gegenstandstheorie und Psychologie* (Leipzig: Barth, 1904).

If no one thought about Hamlet, there would be nothing left to him; if no one had thought about Napoleon, he would have soon seen to it that someone did. The sense of reality is vital in logic, and whoever juggles with it by pretending that Hamlet has another kind of reality is doing a disservice to thought. A robust sense of reality is very necessary in framing a correct analysis of propositions about unicorns, golden mountains, round squares, and other such pseudo-objects.

In obedience to the feeling of reality, we shall insist that, in the analysis of propositions, nothing 'unreal' is to be admitted. But, after all, if there *is* nothing unreal, how, it may be asked, *could* we admit anything unreal? The reply is that, in dealing with propositions, we are dealing in the first instance with symbols, and if we attribute significance to groups of symbols which have no significance, we shall fall into the error of admitting unrealities, in the only sense in which this is possible, namely, as objects described. In the proposition 'I met a unicorn', the whole four words together make a significant proposition, and the word 'unicorn' by itself is significant, in just the same sense as the word 'man'. But the *two* words 'a unicorn' do not form a subordinate group having a meaning of its own. Thus if we falsely attribute meaning to these two words, we find ourselves saddled with 'a unicorn', and with the problem how there can be such a thing in a world where there are no unicorns. 'A unicorn' is an indefinite description which describes nothing. It is not an indefinite description which describes something unreal. Such a proposition as 'x is unreal' only has meaning when 'x' is a description, definite or indefinite; in that case the proposition will be true if 'x' is a description which describes nothing. But whether the description 'x' describes something or describes nothing, it is in any case not a constituent of the proposition in which it occurs; like 'a unicorn' just now, it is not a subordinate group having a meaning of its own. All this results from the fact that, when 'x' is a description, 'x is unreal' or 'x does not exist' is not nonsense, but is always significant and sometimes true.

We may now proceed to define generally the meaning of propositions which contain ambiguous descriptions. Suppose we wish to make some statement about 'a so-and-so', where 'so-and-so's' are those objects that have a certain property ϕ, i.e. those objects x for which the propositional function ϕx is true (e.g. if we take 'a man' as our instance of 'a so-and-so', ϕx will be 'x is human'). Let us now wish to assert the property ψ of 'a so-and-so', i.e. we wish to assert that 'a so-and-so' has that property which x has when ψx is true (e.g. in the case of 'I met a man', ψx will be 'I met x'). Now the proposition that 'a so-and-so' has the property ψ is *not* a proposition of the form 'ψx'. If it were, 'a-so-and-so' would have to be identical with x for a suitable x; and although (in a sense) this may be true in some cases, it is certainly not true in such a case as 'a unicorn'. It is just this fact, that the statement that a so-and-so has the property ψ is not of the form ψx, which makes it possible for 'a so-and-so' to be, in a certain clearly definable sense, 'unreal'. The definition is as follows:

The statement that 'an object having the property ϕ has the property ψ'

means:

'The joint assertion of ϕx and ψx is not always false.'

So far as logic goes, this is the same proposition as might be expressed by 'some ϕ's are ψ's'; but rhetorically there is a difference, because in the one case there is a suggestion of singularity, and in the other case of plurality. This, however, is not the important point. The important point is that, when rightly analysed, propositions verbally about 'a so-and-so' are found to contain no constituent represented by this phrase. And that is why such propositions can be significant even when there is no such thing as a so-and-so.

The definition of *existence*, as applied to ambiguous descriptions, results from what was said at the end of the preceding chapter. We say that 'men exist' or 'a man exists' if the propositional function 'x is human' is sometimes true; and generally 'a so-and-so' exists if 'x is so-and-so' is sometimes true. We may put this in other language. The proposition 'Socrates is a man' is no doubt *equivalent* to 'Socrates is human', but it is not the very same proposition. The *is* of 'Socrates is human' expresses the relation of subject and predicate; the *is* of 'Socrates is a man' expresses identity. It is a disgrace to the human race that it has chosen to employ the same word 'is' for these two entirely different ideas—a disgrace which a symbolic logical language of course remedies. The identity in 'Socrates is a man' is identity between an object named (accepting 'Socrates' as a name, subject to qualifications explained later) and an object ambiguously described. An object ambiguously described will 'exist' when at least one such proposition is true, i.e. when there is at least one true proposition of the form 'x is a so-and-so', where 'x' is a name. It is characteristic of ambiguous (as opposed to definite) descriptions that there may be any number of true propositions of the above form—Socrates is a man, Plato is a man, etc. Thus 'a man exists' follows from Socrates, or Plato, or anyone else. With definite descriptions, on the other hand, the corresponding form of proposition, namely, 'x is the so-and-so' (where 'x' is a name), can only be true for one value of x at most. This brings us to the subject of definite descriptions, which are to be defined in a way analogous to that employed for ambiguous descriptions, but rather more complicated.

We come now to the main subject of the present chapter, namely, the definition of the word *the* (in the singular). One very important point about the definition of 'a so-and-so' applies equally to 'the so-and-so'; the definition to be sought is a definition of propositions in which this phrase occurs, not a definition of the phrase itself in isolation. In the case of 'a so-and-so', this is fairly obvious: no one could suppose that 'a man' was a definite object, which could be defined by itself. Socrates is a man, Plato is a man, Aristotle is a man, but we cannot infer that 'a man' means the same as 'Socrates' means and also the same as 'Plato' means and also the same as 'Aristotle' means, since these three names have different meanings. Nevertheless, when we have enumerated all the men in the world, there is nothing left of which we can say, 'This is a man, and not only so, but it is *the* "a man", the quintessential entity that is just an indefinite man without being anybody in particular.' It is of course quite clear that whatever there is in the world is definite: if it is a man it is one definite man and not any other. Thus there cannot be such an entity as 'a man' to be found in the world, as opposed to specific men. And accordingly it is natural that we do not define 'a man' itself, but only the propositions in which it occurs.

In the case of 'the so-and-so' this is equally true, though at first sight less obvious. We may demonstrate that this must be the case, by a consideration of the difference between a *name* and a *definite description*. Take the proposition, 'Scott is the author of *Waverley*.' We have here a name, 'Scott', and a description, 'the author of *Waverley*', which are asserted to apply to the same person. The distinction between a name and all other symbols may be explained as follows:

A name is a simple symbol whose meaning is something that can only occur as subject, i.e. something of the kind that, in Chapter XIII, we defined as an 'individual' or a 'particular'. And a 'simple' symbol is one which has no parts that are symbols. Thus 'Scott' is a simple symbol, because, though it has parts (namely, separate letters), these parts are not symbols. On the other hand, 'the author of *Waverley*' is not a simple symbol, because the separate words that compose the phrase are parts which are symbols. If, as may be the case, whatever *seems* to be an 'individual' is really capable of further analysis, we shall have to content ourselves with what may be called 'relative individuals', which will be terms that, throughout the context in question, are never analysed and never occur otherwise than as subjects. And in that case we shall have correspondingly to content ourselves with 'relative names'. From the standpoint of our present problem, namely, the definition of descriptions, this problem, whether these are absolute names or only relative names, may be ignored, since it concerns different stages in the hierarchy of 'types', whereas we have to compare such couples as 'Scott' and 'the author of *Waverley*', which both apply to the same object, and do not raise the problem of types. We may, therefore, for the moment, treat names as capable of being absolute; nothing that we shall have to say will depend upon this assumption, but the wording may be a little shortened by it.

We have, then, two things to compare: (1) a *name*, which is a simple symbol, directly designating an individual which is its meaning, and having this meaning in its own right, independently of the meanings of all other words; (2) a *description*, which consists of several words, whose meanings are already fixed, and from which results whatever is to be taken as the 'meaning' of the description.

A proposition containing a description is not identical with what that proposition becomes when a name is substituted, even if the name names the same object as the description describes. 'Scott is the author of *Waverley*' is obviously a different proposition from 'Scott is Scott': the first is a fact in literary history, the second a trivial truism. And if we put anyone other than Scott in place of 'the author of *Waverley*', our proposition would become false, and would therefore certainly no longer be the same proposition. But, it may be said, our proposition is essentially of the same form as (say) 'Scott is Sir Walter', in which two names are said to apply to the same person. The reply is that, if 'Scott is Sir Walter' really means 'the person named "Scott" is the person named "Sir Walter"', then the names are being used as descriptions: i.e. the individual, instead of being named, is being described as the person having that name. This is a way in which names are frequently used in practice, and there will, as a rule, be nothing in the phraseology to show whether they are being used in this way or *as* names. When a name is used directly, merely to indicate what we are speaking about, it is no part of the *fact* asserted, or of the falsehood if our assertion happens to be false: it is merely part of the symbolism by which we express our thought. What we want to express is something which might (for example) be translated into a foreign language; it is something for which the actual words are a vehicle, but of which they are no part. On the other hand, when we make a proposition about 'the person called "Scott"', the actual name 'Scott' enters into what we are asserting, and not merely into the language used in making the assertion. Our proposition will now be a different one if we substitute 'the person called "Sir Walter"'. But so long as we are using names *as* names, whether we say 'Scott' or whether we say 'Sir Walter' is as irrelevant to what we are asserting as whether we speak English or French. Thus so long as names are used *as* names, 'Scott is Sir Walter' is the same trivial proposition as 'Scott is Scott.' This completes the proof that 'Scott is the author of *Waverley*' is not the same proposition as results from substituting a name for 'the author of *Waverley*', no matter what name may be substituted.

When we use a variable, and speak of a propositional function, ϕx say, the process of applying general statements about x to particular cases will consist in substituting a name for the letter 'x', assuming that ϕ is a function which has individuals for its arguments. Suppose, for example, that ϕx is 'always true'; let it be, say, the 'law of identity', $x = x$. Then we may substitute for 'x' any name we choose, and we shall obtain a true proposition. Assuming for the moment that 'Socrates', 'Plato', and 'Aristotle' are names (a very rash assumption), we can infer from the law of identity that Socrates is Socrates, Plato is Plato, and Aristotle is Aristotle. But we shall commit a fallacy if we attempt to infer, without further premisses, that the author of *Waverley* is the author of *Waverley*. This results from what we have just proved, that, if we substitute a name for 'the author of *Waverley*' in a proposition, the proposition we obtain is a different one. That is to say, applying the result to our present case: If 'x' is a name, '$x = x$' is not the same proposition as 'the author of *Waverley* is the author of *Waverley*', no matter what name 'x' may be. Thus from the fact that all propositions of the form '$x = x$' are true we cannot infer, without more ado, that the author of *Waverley* is the author of *Waverley*. In fact, propositions of the form 'the so-and-so is the so-and-so' are not always true: it is necessary that the so-and-so should *exist* (a term which will be explained shortly). It is false that the present King of France is the present King of France, or that the round square is the round square. When we substitute a description for a name, propositional functions which are 'always true' may become false, if the description describes nothing. There is no mystery in this as soon as we realize (what was proved in the preceding paragraph) that when we substitute a description the result is not a value of the propositional function in question.

We are now in a position to define propositions in which a definite description occurs. The only thing that distinguishes 'the so-and-so' from 'a so-and-so' is the implication of uniqueness. We cannot speak of '*the* inhabitant of London', because inhabiting London is an attribute which is not unique. We cannot speak about 'the present King of France', because there is none; but we can speak about 'the present King of England'. Thus propositions about 'the so-and-so' always imply the corresponding propositions about 'a so-and-so', with the addendum that there is not more than one so-and-so. Such a proposition as 'Scott is the author of Waverley' could not be true if *Waverley* had never been written, or if several people had written it; and no more could any other proposition resulting from a propositional function x by the substitution of 'the author of *Waverley*' for 'x'. We may say that 'the author of *Waverley*' means 'the value of x for which "x wrote *Waverley*" is true'. Thus the proposition 'the author of *Waverley* was Scotch', for example, involves:

(1) 'x wrote *Waverley*' is not always false;
(2) 'if x and y wrote *Waverley*, x and y are identical' is always true;
(3) 'if x wrote *Waverley*, x was Scotch' is always true.

These three propositions, translated into ordinary language, state:

(1) at least one person wrote *Waverley*;
(2) at most one person wrote *Waverley*;
(3) whoever wrote *Waverley* was Scotch.

All these three are implied by 'the author of *Waverley* was Scotch.' Conversely, the three together (but no two of them) imply that the author of *Waverley* was Scotch. Hence the three together may be taken as defining what is meant by the proposition 'the author of *Waverley* was Scotch.'

We may somewhat simplify these three propositions. The first and second together are equivalent to: 'There is a term c such that "x wrote *Waverley*" is true when x is c and is false when x is not c.' In other words, 'There is a term c such that "x wrote *Waverley*" is always equivalent to "x is c".' (Two propositions are 'equivalent' when both are true or both are false.) We have here, to begin with, two functions of x, 'x wrote *Waverley*' and 'x is c', and we form a function of c by considering the equivalence of these two functions of x for all values of x; we then proceed to assert that the resulting function of c is 'sometimes true', i.e. that it is true for at least one value of c. (It obviously cannot be true for more than one value of c.) These two conditions together are defined as giving the meaning of 'the author of *Waverley* exists.'

We may now define 'the term satisfying the function ϕx exists.' This is the general form of which the above is a particular case. 'The author of *Waverley*' is 'the term satisfying the function "x wrote *Waverley*".' And 'the so-and-so' will always involve reference to some propositional function, namely, that which defines the property that makes a thing a so-and-so. Our definition is as follows:

'The term satisfying the function ϕx exists' means:
'There is a term c such that ϕx is always equivalent to "x is c".'

In order to define 'the author of *Waverley* was Scotch', we have still to take account of the third of our three propositions, namely, 'Whoever wrote *Waverley* was Scotch.' This will be satisfied by merely adding that the c in question is to be Scotch. Thus 'the author of *Waverley* was Scotch' is:

'There is a term c such that (1) "x wrote *Waverley*" is always equivalent to "x is c", (2) c is Scotch.'

And generally: 'the term satisfying ϕx satisfies ψx' is defined as meaning:

'There is a term c such that (1) ϕx is always equivalent to "x is c", (2) ψc is true.'

This is the definition of propositions in which descriptions occur.

It is possible to have much knowledge concerning a term described, i.e. to know many propositions concerning 'the so-and-so', without actually knowing what the so-and-so is, i.e. without knowing any proposition of the form 'x is the so-and-so', where 'x' is a name. In a detective story propositions about 'the man who did the deed' are accumulated, in the hope that ultimately they will suffice to demonstrate that it was A who did the deed. We may even go so far as to say that, in all such knowledge as can be expressed in words—with the exception of 'this' and 'that' and a few other words of which the meaning varies on different occasions—no names, in the strict sense, occur, but what seem like names are really descriptions. We may inquire significantly whether Homer existed, which we could not do if 'Homer' were a name. The proposition 'the so-and-so exists' is significant, whether true or false; but if a is the so-and-so (where 'a' is a name), the words 'a exists' are meaningless. It is only of descriptions—definite or indefinite—that existence can be significantly asserted; for, if 'a' is a name, it *must* name something: what does not name anything is not a name, and therefore, if intended to be a name, is a symbol devoid of meaning, whereas a description, like 'the present King of France', does not become incapable of occurring significantly merely on the ground that it describes nothing, the reason being that it is a *complex* symbol, of which the meaning is derived from that of its constituent symbols. And so, when we ask whether Homer existed, we are using the word 'Homer' as an abbreviated description: we may replace it by (say) 'the author of the *Iliad* and the

Odyssey'. The same considerations apply to almost all uses of what look like proper names.

When descriptions occur in propositions, it is necessary to distinguish what may be called 'primary' and 'secondary' occurrences. The abstract distinction is as follows. A description has a 'primary' occurrence when the proposition in which it occurs results from substituting the description for 'x' in some propositional function ϕx; a description has a 'secondary' occurrence when the result of substituting the description for x in ϕx gives only *part* of the proposition concerned. An instance will make this clearer. Consider 'the present King of France is bald.' Here 'the present King of France' has a primary occurrence, and the proposition is false. Every proposition in which a description which describes nothing has a primary occurrence is false. But now consider 'the present King of France is not bald.' This is ambiguous. If we are first to take 'x is bald', then substitute 'the present King of France' for 'x', and then deny the result, the occurrence of 'the present King of France' is secondary and our proposition is true; but if we are to take 'x is not bald' and substitute 'the present King of France' for 'x', then 'the present King of France' has a primary occurrence and the proposition is false. Confusion of primary and secondary occurrences is a ready source of fallacies where descriptions are concerned.

IV

ON REFERRING

P. F. STRAWSON

I

We very commonly use expressions of certain kinds to mention or refer to some individual person or single object or particular event or place or process, in the course of doing what we should normally describe as making a statement about that person, object, place, event, or process. I shall call this way of using expressions the 'uniquely referring use'. The classes of expressions which are most commonly used in this way are: singular demonstrative pronouns ('this' and 'that'); proper names (e.g. 'Venice', 'Napoleon', 'John'); singular personal and impersonal pronouns ('he', 'she', 'I', 'you', 'it'); and phrases beginning with the definite article followed by a noun, qualified or unqualified, in the singular (e.g. 'the table', 'the old man', 'the king of France'). Any expression of any of these classes can occur as the subject of what would traditionally be regarded as a singular subject-predicate sentence; and would, so occurring, exemplify the use I wish to discuss.

I do not want to say that expressions belonging to these classes never have any other use than the one I want to discuss. On the contrary, it is obvious that they do. It is obvious that anyone who uttered the sentence, 'The whale is a mammal', would be using the expression 'the whale' in a way quite different from the way it would be used by anyone who had occasion seriously to utter the sentence, 'The whale struck the ship'. In the first sentence one is obviously *not* mentioning, and in the second sentence one obviously *is* mentioning, a particular whale. Again if I said, 'Napoleon was the greatest French soldier', I should be using the word 'Napoleon' to mention a certain individual, but I should not be using the phrase, 'the greatest French soldier', to mention an individual, but to say something about an individual I had already mentioned. It would be natural to say that in using this sentence I was talking *about* Napoleon and that what I was *saying* about him was that he was the greatest French soldier. But of course I *could* use the expression, 'the greatest

From *Mind*, 59 (1950), 320–44. Reprinted by permission of Oxford University Press. Note 2 was added for the reprint in G. H. R. Parkinson (ed.), *The Theory of Meaning* (Oxford: Oxford University Press, 1968).

French soldier', to mention an individual; for example, by saying: 'The greatest French soldier died in exile.' So it is obvious that at least some expressions belonging to the classes I mentioned *can* have uses other than the use I am anxious to discuss. Another thing I do not want to say is that in any given sentence there is never more than one expression used in the way I propose to discuss. On the contrary, it is obvious that there may be more than one. For example, it would be natural to say that, in seriously using the sentence, 'The whale struck the ship', I was saying something about both a certain whale and a certain ship, that I was using each of the expressions 'the whale' and 'the ship' to mention a particular object; or, in other words, that I was using each of these expressions in the uniquely referring way. In general, however, I shall confine my attention to cases where an expression used in this way occurs as the grammatical subject of a sentence.

I think it is true to say that Russell's Theory of Descriptions, which is concerned with the last of the four classes of expressions I mentioned above (i.e. with expressions of the form 'the so-and-so') is still widely accepted among logicians as giving a correct account of the use of such expressions in ordinary language. I want to show, in the first place, that this theory, so regarded, embodies some fundamental mistakes.

What question or questions about phrases of the form 'the so-and-so' was the Theory of Descriptions designed to answer? I think that at least one of the questions may be illustrated as follows. Suppose someone were now to utter the sentence, 'The king of France is wise.' No one would say that the sentence which had been uttered was meaningless. Everyone would agree that it was significant. But everyone knows that there is not at present a king of France. One of the questions the Theory of Descriptions was designed to answer was the question: how can such a sentence as 'The king of France is wise' be significant even when there is nothing which answers to the description it contains, i.e., in this case, nothing which answers to the description 'The king of France'? And one of the reasons why Russell thought it important to give a correct answer to this question was that he thought it important to show that another answer which might be given was wrong. The answer that he thought was wrong, and to which he was anxious to supply an alternative, might be exhibited as the conclusion of either of the following two fallacious arguments. Let us call the sentence 'The king of France is wise' the sentence S. Then the first argument is as follows:

(1) The phrase, 'the king of France', is the subject of the sentence S. Therefore (2) if S is a significant sentence, S is a sentence *about* the king of France.

But (3) if there in no sense exists a king of France, the sentence is not about anything, and hence not about the king of France.
Therefore (4) since S is significant, there must in some sense (in some world) exist (or subsist) the king of France.

And the second argument is as follows:

(1) If S is significant, it is either true or false.
(2) S is true if the king of France is wise and false if the king of France is not wise.
(3) But the statement that the king of France is wise and the statement that the king of France is not wise are alike true only if there is (in some sense, in some world) something which is the king of France.
Hence (4) since S is significant, there follows the same conclusion as before.

These are fairly obviously bad arguments, and, as we should expect, Russell rejects them. The postulation of a world of strange entities, to which the king of France belongs, offends, he says, against 'that feeling for reality which ought to be preserved even in the most abstract studies'. The fact that Russell rejects these arguments is, however, less interesting than the extent to which, in rejecting their conclusion, he concedes the more important of their principles. Let me refer to the phrase, 'the king of France', as the phrase D. Then I think Russell's reasons for rejecting these two arguments can be summarized as follows. The mistake arises, he says, from thinking that D, which is certainly the *grammatical* subject of S, is also the *logical* subject of S. But D is not the logical subject of S. In fact S, although grammatically it has a singular subject and a predicate, is not logically a subject-predicate sentence at all. The proposition it expresses is a complex kind of *existential* proposition, part of which might be described as a 'uniquely existential' proposition. To exhibit the logical form of the proposition, we should rewrite the sentence in a logically appropriate grammatical form; in such a way that the deceptive similarity of S to a sentence expressing a subject-predicate proposition would disappear, and we should be safeguarded against arguments such as the bad ones I outlined above. Before recalling the details of Russell's analysis of S, let us notice what his answer, as I have so far given it, seems to imply. His answer seems to imply that in the case of a sentence which is similar to S in that (1) it is grammatically of the subject-predicate form and (2) its grammatical subject does not refer to anything, then the only alternative to its being meaningless is that it should not really (i.e. logically) be of the subject-predicate form at all, but of some quite different form. And this in its turn seems to imply that if there are any sentences which are genuinely of the subject-predicate form, then the very fact of their being significant, having a meaning, guarantees that there *is* something referred to by the logical (and grammatical) subject. Moreover, Russell's answer seems to imply that there are such sentences. For if it is true that one may be misled by the grammatical similarity of S to other sentences into thinking that it is logically of the subject-predicate form, then surely there must be other sentences grammatically similar to S, which *are* of the subject-predicate form. To show not only that Russell's answer seems to imply these conclusions, but that he accepted at least the first two of them, it is enough to consider what he says about a class of expressions which he calls 'logically proper names' and contrasts with expressions, like D, which he calls 'definite descriptions'. Of logically proper names Russell says or implies the following things:

(1) That they and they alone can occur as subjects of sentences which are genuinely of the subject-predicate form;

(2) that an expression intended to be a logically proper name is *meaningless* unless there is some single object for which it stands: for the *meaning* of such an expression just is the individual object which the expression designates. To be a name at all, therefore, it *must* designate something.

It is easy to see that if anyone believes these two propositions, then the only way for him to save the significance of the sentence S is to deny that it is a logically subject-predicate sentence. Generally, we may say that Russell recognizes only two ways in which sentences which seem, from their grammatical structure, to be about some particular person or individual object or event, can be significant:

(1) The first is that their grammatical form should be misleading as to their logical form, and that they should be analysable, like S, as a special kind of existential sentence;

(2) The second is that their grammatical subject should be a logically proper name, of which the meaning is the individual thing it designates.

I think that Russell is unquestionably wrong in this, and that sentences which are significant, and which begin with an expression used in the uniquely referring way, fall into neither of these two classes. Expressions used in the uniquely referring way are never either logically proper names or descriptions, if what is meant by calling them 'descriptions' is that they are to be analysed in accordance with the model provided by Russell's Theory of Descriptions.

There are no logically proper names and there are no descriptions (in this sense).

Let us now consider the details of Russell's analysis. According to Russell, anyone who asserted S would be asserting that:

(1) There is a king of France.
(2) There is not more than one king of France.
(3) There is nothing which is king of France and is not wise.

It is easy to see both how Russell arrived at this analysis, and how it enables him to answer the question with which we began, viz. the question: How can the sentence S be significant when there is no king of France? The way in which he arrived at the analysis was clearly by asking himself what would be the circumstances in which we would say that anyone who uttered the sentence S had made a true assertion. And it does seem pretty clear, and I have no wish to dispute, that the sentences (1)–(3) above do describe circumstances which are at least *necessary* conditions of anyone making a true assertion by uttering the sentence S. But, as I hope to show, to say this is not at all the same thing as to say that Russell has given a correct account of the use of the sentence S or even that he has given an account which, though incomplete, is correct as far as it goes; and is certainly not at all the same thing as to say that the model translation provided is a correct model for all (or for any) singular sentences beginning with a phrase of the form 'the so-and-so'.

It is also easy to see how this analysis enables Russell to answer the question of how the sentence S can be significant, even when there is no king of France. For, if this analysis is correct, anyone who utters the sentence S to-day would be jointly asserting three propositions, one of which (viz. that there is a king of France) would be false; and since the conjunction of three propositions, of which one is false, is itself false, the assertion as a whole would be significant, but false. So neither of the bad arguments for subsistent entities would apply to such an assertion.

II

As a step towards showing that Russell's solution of his problem is mistaken, and towards providing the correct solution, I want now to draw certain distinctions. For this purpose I shall, for the remainder of this section, refer to an expression which has a uniquely referring use as 'an expression' for short; and to a sentence beginning with such an expression as 'a sentence' for short. The distinctions I shall draw are rather rough and ready, and, no doubt, difficult cases could be produced which would call for their refinement. But I think they will serve my purpose. The distinctions are between:

(A1) a sentence,
(A2) a use of a sentence,
(A3) an utterance of a sentence,

and, correspondingly, between:

(B1) an expression,
(B2) a use of an expression,
(B3) an utterance of an expression.

Consider again the sentence, 'The king of France is wise.' It is easy to imagine that this sentence was uttered at various times from, say, the beginning of the seventeenth century onwards, during the reigns of each successive French monarch; and easy to imagine that it was also uttered during the subsequent periods in which France was not a monarchy. Notice that it was natural for me to speak of 'the sentence' or 'this sentence' being uttered at various times during this period; or, in other words, that it would be natural and correct to speak of *one and the same* sentence being uttered on all these various occasions. It is in the sense in which it would be correct to speak of one and the same sentence being uttered on all these various occasions that I want to use the expression (A1) 'a sentence'. There are, however, obvious differences between different *occasions of the use* of this sentence. For instance, if one man uttered it in the reign of Louis XIV and another man uttered it in the reign of Louis XV, it would be natural to say (to assume) that they were respectively talking about different people; and it might be held that the first man, in using the sentence, made a true assertion, while the second man, in using the same sentence, made a false assertion. If on the other hand two different men simultaneously uttered the sentence (e.g. if one wrote it and the other spoke it) during the reign of Louis XIV, it would be natural to say (assume) that they were both talking about the same person, and, in that case, in using the sentence, they *must* either both have made a true assertion or both have made a false assertion. And this illustrates what I mean by *a use* of a sentence. The two men who uttered the sentence, one in the reign of Louis XV and one in the reign of Louis XIV, each made a different use of the same sentence; whereas the two men who uttered the sentence simultaneously in the reign of Louis XIV, made the same use[1] of the same sentence. Obviously in the case of this sentence, and equally obviously in the case of many others, we cannot

[1] This usage of 'use' is, of course, different from (*a*) the current usage in which 'use' (of a particular word, phrase, sentence) = (roughly) 'rules for using' = (roughly) 'meaning'; and from (*b*) my own usage in the phrase 'uniquely referring use of expressions' in which 'use' = (roughly) 'way of using'.

talk of *the sentence* being true or false, but only of its being used to make a true or false assertion, or (if this is preferred) to express a true or a false proposition. And equally obviously we cannot talk of *the sentence* being *about* a particular person, for the same sentence may be used at different times to talk about quite different particular persons, but only of *a use* of the sentence to talk about a particular person. Finally it will make sufficiently clear what I mean by an utterance of a sentence if I say that the two men who simultaneously uttered the sentence in the reign of Louis XIV made two different utterances of the same sentence, though they made the same *use* of the sentence.

If we now consider not the whole sentence, 'The king of France is wise', but that part of it which is the expression, 'the king of France', it is obvious that we can make analogous, though not identical distinctions between (1) the expression, (2) a use of the expression and (3) an utterance of the expression. The distinctions will not be identical; we obviously cannot correctly talk of the expression 'the king of France' being used to express a true or false proposition, since in general only sentences can be used truly or falsely; and similarly it is only by using a sentence and not by using an expression alone, that you can talk about a particular person. Instead, we shall say in this case that you *use* the expression to *mention* or *refer to* a particular person in the course of using the sentence to talk about him. But obviously in this case, and a great many others, the *expression* (B1) cannot be said to mention, or refer to, anything, any more than the *sentence* can be said to be true or false. The same expression can have different mentioning-uses, as the same sentence can be used to make statements with different truth-values. 'Mentioning', or 'referring', is not something an expression does; it is something that someone can use an expression to do. Mentioning, or referring to, something is a characteristic of *a use* of an expression, just as 'being about' something, and truth-or-falsity, are characteristics of *a use* of a sentence.

A very different example may help to make these distinctions clearer. Consider another case of an expression which has a uniquely referring use, viz. the expression 'I'; and consider the sentence, 'I am hot'. Countless people may use this same sentence; but it is logically impossible for two different people to make *the same use* of this sentence; or, if this is preferred, to use it to express the same proposition. The expression 'I' may correctly be used by (and only by) any one of innumerable people to refer to himself. To say this is to say something about the expression 'I': it is, in a sense, to give its meaning. This is the sort of thing that can be said about *expressions*. But it makes no sense to say of the *expression*

'I' that it refers to a particular person. This is the sort of thing that can be said only of a particular use of the expression.

Let me use 'type' as an abbreviation for 'sentence or expression'. Then I am not saying that there are sentences and expressions (types), *and* uses of them, *and* utterances of them, as there are ships *and* shoes *and* sealing-wax. I am saying that we cannot say *the same things* about types, uses of types, and utterances of types. And the fact is that we do talk about types; and that confusion is apt to result from the failure to notice the differences between what we can say about these and what we can say only about the *uses* of types. We are apt to fancy we are talking about sentences and expressions when we are talking about the uses of sentences and expressions.

This is what Russell does. Generally, as against Russell, I shall say this. Meaning (in at least one important sense) is a function of the sentence or expression; mentioning and referring and truth or falsity, are functions of the use of the sentence or expression. To give the meaning of an expression (in the sense in which I am using the word) is to give *general directions* for its use to refer to or mention particular objects or persons; to give the meaning of a sentence is to give *general directions* for its use in making true or false assertions. It is not to talk about any particular occasion of the use of the sentence or expression. The meaning of an expression cannot be identified with the object it is used, on a particular occasion, to refer to. The meaning of a sentence cannot be identified with the assertion it is used, on a particular occasion, to make. For to talk about the meaning of an expression or sentence is not to talk about its use on a particular occasion, but about the rules, habits, conventions governing its correct use, on all occasions, to refer or to assert. So the question of whether a sentence or expression *is significant or not* has nothing whatever to do with the question of whether the sentence, *uttered on a particular occasion*, is, on that occasion, being used to make a true-or-false assertion or not, or of whether the expression is, on that occasion, being used to refer to, or mention, anything at all.

The source of Russell's mistake was that he thought that referring or mentioning, if it occurred at all, must be meaning. He did not distinguish B1 from B2; he confused expressions with their use in a particular context; and so confused meaning with mentioning, with referring. If I talk about my handkerchief, I can, perhaps, produce the object I am referring to out of my pocket. I can't produce the meaning of the expression, 'my handkerchief', out of my pocket. Because Russell confused meaning with mentioning, he thought that if there were any expressions having a uniquely referring use, which were what they seemed (i.e. logical subjects) and not something else in disguise, their meaning must *be* the particular object which they were used to refer to. Hence the troublesome mythology of the logically proper name. But if someone asks me the meaning of the expression 'this'—once Russell's favourite candidate for this status—I do not hand him the object I have just used the expression to refer to, adding at the same time that the meaning of the word changes every time it is used. Nor do I hand him all the objects it ever has been, or might be, used to refer to. I explain and illustrate the conventions governing the use of the expression. This *is* giving the meaning of the expression. It is quite different from giving (in any sense of giving) the object to which it refers; for the expression itself does not refer to anything; though it can be used, on different occasions, to refer to innumerable things. Now as a matter of fact there is, in English, a sense of the word 'mean' in which this word does approximate to 'indicate, mention or refer to'; e.g. when somebody (unpleasantly) says, 'I mean you'; or when I point and say, 'That's the one I mean'. But *the one I meant* is quite different from *the meaning of the expression* I used to talk of it. In this special sense of 'mean', it is people who mean, not expressions. People use expressions to refer to particular things. But the meaning of an expression is not the set of things or the single thing it may correctly be used to refer to: the meaning is the set of rules, habits, conventions for its use in referring.

It is the same with sentences: even more obviously so. Everyone knows that the sentence, 'The table is covered with books', is significant, and everyone knows what it means. But if I ask, 'What object is that sentence about?' I am asking an absurd question—a question which cannot be asked about the sentence, but only about some use of the sentence: and in this case the sentence hasn't been used, it has only been taken as an example. In knowing what it means, you are knowing how it could correctly be used to talk about things: so knowing the meaning hasn't anything to do with knowing about any particular use of the sentence to talk about anything. Similarly, if I ask: 'Is the sentence true or false?' I am asking an absurd question, which becomes no less absurd if I add, 'It must be one or the other since it's significant.' The question is absurd, because the *sentence* is neither true nor false any more than it's *about* some object. Of course the fact that it's significant is the same as the fact that it *can* correctly be used to talk about something and that, in so using it, someone will be making a true or false assertion. And I will add that it will be used to make a true or false assertion *only* if the person using it *is* talking about something. If, when he utters it, he is not talking about anything, then his use is not a genuine one, but a spurious or pseudo-use: he is not making either a true or a false assertion, though he may think he is. And this points the way to the correct answer to the puzzle to which the Theory of Descriptions gives a fatally incorrect answer. The important point is that the question of whether the sentence is significant or not is quite independent of the question that can be raised about a particular use of it, viz. the question whether it is a genuine or a spurious use, whether it is being used to talk about something, or in make-believe, or as an example in philosophy. The question whether the sentence is significant or not is the question whether there exist such language habits, conventions or rules that the sentence logically could be used to talk about something; and is hence quite independent of the question whether it is being so used on a particular occasion.

III

Consider again the sentence, 'The king of France is wise', and the true and false things Russell says about it.

There are at least two true things which Russell would say about the sentence:

(1) The first is that it is significant; that if anyone were now to utter it, he would be uttering a significant sentence.

(2) The second is that anyone now uttering the sentence would be making a true assertion only if there in fact at present existed one and only one king of France, and if he were wise.

What are the false things which Russell would say about the sentence? They are:

(1) That anyone now uttering it would be making a true assertion or a false assertion;

(2) That part of what he would be asserting would be that there at present existed one and only one king of France.

I have already given some reasons for thinking that these two statements are incorrect. Now suppose someone were in fact to say to you with a perfectly serious air: 'The king of France is wise.' Would you say, 'That's untrue'? I think it's quite certain that you wouldn't. But suppose he went on to *ask* you whether you thought that what he had just said was true, or was false; whether you agreed or disagreed with what he had just said. I think you would be inclined, with some hesitation, to say that you didn't do either; that the question of whether his statement was true or false simply *didn't arise*, because there was no such person as the king of France.[2] You might, if he were obviously serious (had a dazed astray-in-the-centuries look), say something like: 'I'm afraid you must be under a misapprehension. France is not a monarchy. There is no king of France.' And this brings out the point that if a man seriously uttered the sentence, his uttering it would in some sense be *evidence* that he *believed* that there was a king of France. It would not be evidence for his believing this simply in the way in which a man's reaching for his raincoat is evidence for his believing that it is raining. But nor would it be evidence for his believing this in the way in which a man's saying, 'It's raining' is evidence for his believing that it is raining. We might put it as follows. To say, 'The king of France is wise' is, in some sense of 'imply', to *imply* that there is a king of France. But this is a very special and odd sense of 'imply'. 'Implies' in this sense is certainly not equivalent to 'entails' (or 'logically implies'). And this comes out from the fact that when, in response to his statement, we say (as we should) 'There is no king of France', we should certainly *not* say we were *contradicting* the statement that the king of France is wise. We are certainly not saying that it's false. We are, rather, giving a reason for saying that the question of whether it's true or false simply doesn't arise.

And this is where the distinction I drew earlier can help us. The sentence, 'The king of France is wise', is certainly significant; but this does not mean that any particular use of it is true or false. We use it truly or falsely when we use it to talk about someone; when, in using the expression, 'The king of France', we are in fact mentioning someone. The fact that the sentence and the expression, respectively, are significant just is the fact that the sentence *could* be used, in certain circumstances, to say something true or false, that the expression *could* be used, in certain circumstances, to mention a particular person; and to know their meaning is to know what sort of circumstances these are. So when we utter the sentence without in fact mentioning anybody by the use of the phrase, 'The king of France', the sentence doesn't cease to be significant: we simply *fail* to say anything true or false because we simply fail to mention anybody by this particular use of that perfectly significant phrase. It is, if you like, a spurious use of the sentence, and a spurious use of the expression; though we may (or may not) mistakenly think it a genuine use.

And such spurious uses are very familiar. Sophisticated romancing,

[2] Since this article was written, there has appeared a clear statement of this point by Mr Geach in *Analysis*, 10: 4 (March 1950), 84–8.

sophisticated fiction,[3] depend upon them. If I began, 'The king of France is wise', and went on, 'and he lives in a golden castle and has a hundred wives', and so on, a hearer would understand me perfectly well, without supposing *either* that I was talking about a particular person, *or* that I was making a false statement to the effect that there existed such a person as my words described. (It is worth adding that where the use of sentences and expressions is overtly fictional, the sense of the word 'about' may change. As Moore said, it is perfectly natural and correct to say that some of the statements in *Pickwick Papers* are *about* Mr Pickwick. But where the use of sentences and expressions is not overtly fictional, this use of 'about' seems less correct; i.e. it would not *in general* be correct to say that a statement was about Mr X or the so-and-so, unless there were such a person or thing. So it is where the romancing is in danger of being taken seriously that we might answer the question, 'Who is he talking about?' with 'He's not talking about anybody'; but, in saying this, we are not saying that what he is saying is either false or nonsense.)

Overtly fictional uses apart, however, I said just now that to use such an expression as 'The king of France' at the beginning of a sentence was, in some sense of 'imply', to imply that there was a king of France. When a man uses such an expression, he does not *assert*, nor does what he says *entail*, a uniquely existential proposition. But one of the conventional functions of the definite article is to act as a *signal* that a unique reference is being made—a signal, not a disguised assertion. When we begin a sentence with 'the such-and-such' the use of 'the' shows, but does not state, that we are, or intend to be, referring to one particular individual of the species 'such-and-such'. *Which* particular individual is a matter to be determined from context, time, place and any other features of the situation of utterance. Now, whenever a man uses any expression, the presumption is that he thinks he is using it correctly: so when he uses the expression, 'the such-and-such', in a uniquely referring way, the presumption is that he thinks both that there is *some* individual of that species, and that the context of use will sufficiently determine which one he has in mind. To use the word 'the' in this way is then to imply (in the relevant sense of 'imply') that the existential conditions described by Russell are fulfilled. But to use 'the' in this way is not to *state* that those conditions are fulfilled. If I begin a sentence with an expression of the form, 'the so-and-so', and then am prevented from

[3] The unsophisticated kind begins: 'Once upon time there was . . .'.

saying more, I have made no statement of any kind; but I may have succeeded in mentioning someone or something.

The uniquely existential assertion supposed by Russell to be part of any assertion in which a uniquely referring use is made of an expression of the form 'the so-and-so' is, he observes, a compound of two assertions. To say that there is a ϕ is to say something compatible with there being several ϕs; to say there is not more than one ϕ is to say something compatible with there being none. To say there is one ϕ and one only is to compound these two assertions. I have so far been concerned mostly with the alleged assertion of existence and less with the alleged assertion of uniqueness. An example which throws the emphasis on to the latter will serve to bring out more clearly the sense of 'implied' in which a uniquely existential assertion is implied, but not entailed, by the use of expressions in the uniquely referring way. Consider the sentence, 'The table is covered with books.' It is quite certain that in any normal use of this sentence, the expression 'the table' would be used to make a unique reference, i.e. to refer to some one table. It is a quite strict use of the definite article, in the sense in which Russell talks on p. 30 of *Principia Mathematica*, of using the article '*strictly*, so as to imply uniqueness'. On the same page Russell says that a phrase of the form 'the so-and-so', used strictly, 'will only have an application in the event of there being one so-and-so and no more'. Now it is obviously quite false that the phrase 'the table' in the sentence 'the table is covered with books', used normally, will 'only have an application in the event of there being one table and no more'. It is indeed tautologically true that, in such a use, the phrase will have an application only in the event of there being one table and no more *which is being referred to*, and that it will be understood to have an application only in the event of there being one table and no more which it is understood as being used to refer to. To use the sentence is not to assert, but it is (in the special sense discussed) to imply, that there is only one thing which is *both* of the kind specified (i.e. a table) *and is being referred to* by the speaker. It is obviously not to assert this. To refer is not to say you are referring. To say there is *some table or other* to which you are referring is not the same as referring to a particular table. We should have no use for such phrases as 'the individual I referred to' unless there were something which counted as referring. (It would make no sense to say you had pointed if there were nothing which counted as pointing.) So once more I draw the conclusion that referring to or mentioning a particular thing cannot be dissolved into any kind of assertion. To refer is not to assert, though you refer in order to go on to assert.

Let me now take an example of the uniquely referring use of an expression not of the form, 'the so-and-so'. Suppose I advance my hands, cautiously cupped, towards someone, saying, as I do so, 'This is a fine red one.' He, looking into my hands and seeing nothing there, may say: 'What is? What are you talking about?' Or perhaps, 'But there's nothing in your hands.' Of course it would be absurd to say that in saying 'But you've got nothing in your hands', he was *denying* or *contradicting* what I said. So 'this' is not a disguised description in Russell's sense. Nor is it a logically proper name. For one must know what the sentence means in order to react in that way to the utterance of it. It is precisely because the significance of the word 'this' is independent of any particular reference it may be used to make, though not independent of the way it may be used to refer, that I can, as in this example, use it to *pretend* to be referring to something.

The general moral of all this is that communication is much less a matter of explicit or disguised assertion than logicians used to suppose. The particular application of this general moral in which I am interested is its application to the case of making a unique reference. It is a part of the significance of expressions of the kind I am discussing that they can be used, in an immense variety of contexts, to make unique references. It is no part of their significance to assert that they are being so used or that the conditions of their being so used are fulfilled. So the wholly important distinction we are required to draw is between:

(1) using an expression to make a unique reference; and
(2) asserting that there is one and only one individual which has certain characteristics (e.g. is of a certain kind, or stands in a certain relation to the speaker, or both).

This is, in other words, the distinction between

(1) sentences containing an expression used to indicate or mention or refer to a particular person or thing; and
(2) uniquely existential sentences.

What Russell does is progressively to assimilate more and more sentences of class (1) to sentences of class (2), and consequently to involve himself in insuperable difficulties about logical subjects, and about values for individual variables generally: difficulties which have led him finally to the logically disastrous theory of names developed in the *Enquiry* and in *Human Knowledge*. That view of the meaning of logical-subject-expressions which provides the whole incentive to the Theory of Descriptions at the same time precludes the possibility of Russell's ever finding any satisfactory substitutes for those expressions which, beginning with substantival phrases, he progressively degrades from the status of logical subjects.[4] It is not simply, as is sometimes said, the fascination of the relation between a name and its bearer, that is the root of the trouble. Not even names come up to the impossible standard set. It is rather the combination of two more radical misconceptions: first, the failure to grasp the importance of the distinction (Section II above) between what may be said of an expression and what may be said of a particular use of it; second, a failure to recognize the uniquely referring use of expressions for the harmless, necessary thing it is, distinct from, but complementary to, the predicative or ascriptive use of expressions. The expressions which can in fact occur as singular logical subjects are expressions of the class I listed at the outset (demonstratives, substantival phrases, proper names, pronouns): to say this is to say that these expressions, together with context (in the widest sense) are what one uses to make unique references. The point of the conventions governing the uses of such expressions is, along with the situation of utterance, to secure uniqueness of reference. But to do this, enough is enough. We do not, and we cannot, while referring, attain the point of complete explicitness at which the referring function is no longer performed. The actual unique reference made, if any, is a matter of the particular use in the particular context; the significance of the expression used is the set of rules or conventions which permit such references to be made. Hence we can, using significant expressions, pretend to refer, in make-believe or in fiction, or mistakenly think we are referring when we are not referring to anything.

This shows the need for distinguishing two kinds (among many others) of linguistic conventions or rules: rules for referring, and rules for attributing and ascribing; and for an investigation of the former. If we recognize this distinction of use for what it is, we are on the way to solving a number of ancient logical and metaphysical puzzles.

My last two sections are concerned, but only in the barest outline, with these questions.

IV

One of the main purposes for which we use language is the purpose of stating facts about things and persons and events. If we want to fulfil this purpose, we must have some way of forestalling the question, 'What

[4] And this in spite of the danger-signal of that phrase, '*misleading* grammatical form'.

(who, which one) are you talking about?' as well as the question, 'What are you saying about it (him, her)?' The task of forestalling the first question is the referring (or identifying) task. The task of forestalling the second is the attributive (or descriptive or classificatory or ascriptive) task. In the conventional English sentence which is used to state, or to claim to state, a fact about an individual thing or person or event, the performance of these two tasks can be roughly and approximately assigned to separable expressions.[5] And in such a sentence, this assigning of expressions to their separate roles corresponds to the conventional grammatical classification of subject and predicate. There is nothing sacrosanct about the employment of separable expressions for these two tasks. Other methods could be, and are, employed. There is, for instance, the method of uttering a single word or attributive phrase in the conspicuous presence of the object referred to; or that analogous method exemplified by, e.g. the painting of the words 'unsafe for lorries' on a bridge, or the tying of a label reading 'first prize' on a vegetable marrow. Or one can imagine an elaborate game in which one never used an expression in the uniquely referring way at all, but uttered only uniquely existential sentences, trying to enable the hearer to identify what was being talked of by means of an accumulation of relative clauses. (This description of the purposes of the game shows in what sense it would be a game: this is not the normal use we make of existential sentences.) Two points require emphasis. The first is that the necessity of performing these two tasks in order to state particular facts requires no transcendental explanation: to call attention to it is partly to elucidate the meaning of the phrase, 'stating a fact'. The second is that even this elucidation is made in terms derivative from the grammar of the conventional singular sentence; that even the overtly functional, linguistic distinction between the identifying and attributive roles that words may play in language is prompted by the fact that ordinary speech offers us separable expressions to which the different functions may be plausibly and approximately assigned. And this functional distinction has cast long philosophical shadows. The distinctions between particular and universal, between substance and quality, are such pseudo-material shadows, cast by the grammar of the conventional sentence, in which separable expressions play distinguishable roles.

To use a separate expression to perform the first of these tasks is to use an expression in the uniquely referring way. I want now to say

[5] I neglect relational sentences; for these require, not a modification in the principle of what I say, but a complication of the detail.

something in general about the conventions of use for expressions used in this way, and to contrast them with conventions of ascriptive use. I then proceed to the brief illustration of these general remarks and to some further applications of them.

What in general is required for making a unique reference is, obviously, some device, or devices, for showing both *that* a unique reference is intended and *what* unique reference it is; some device requiring and enabling the hearer or reader to identify what is being talked about. In securing this result, the context of utterance is of an importance which it is almost impossible to exaggerate; and by 'context' I mean, at least, the time, the place, the situation, the identity of the speaker, the subjects which form the immediate focus of interest, and the personal histories of both the speaker and those he is addressing. Besides context, there is, of course, convention—linguistic convention. But, except in the case of genuine proper names, of which I shall have more to say later, the fulfilment of more or less precisely stateable contextual conditions is *conventionally* (or, in a wide sense of the word, *logically*) required for the correct referring use of expressions, in a sense in which this is not true of correct ascriptive uses. The requirement for the correct application of an expression in its ascriptive use to a certain thing is simply that the thing should be of a certain kind, have certain characteristics. The requirement for the correct application of an expression in its referring use to a certain thing is something over and above any requirement derived from such ascriptive meaning as the expression may have; it is, namely, the requirement that the thing should be in a certain relation to the speaker and to the context of utterance. Let me call this the contextual requirement. Thus, for example, in the limiting case of the word 'I' the contextual requirement is that the thing should be identical with the speaker; but in the case of most expressions which have a referring use this requirement cannot be so precisely specified. A further, and perfectly general, difference between conventions for referring and conventions for describing is one we have already encountered, viz. that the fulfilment of the conditions for a correct ascriptive use of an expression is a part of what is stated by such a use; but the fulfilment of the conditions for a correct referring use of an expression is never part of what is stated, though it is (in the relevant sense of 'implied') implied by such a use.

Conventions for referring have been neglected or misinterpreted by logicians. The reasons for this neglect are not hard to see, though they are hard to state briefly. Two of them are, roughly: (1) the preoccupation of most logicians with definitions; (2) the preoccupation of some logicians with formal systems. (1) A definition, in the most familiar sense, is a specification of the conditions of the correct ascriptive or classificatory use of an expression. Definitions take no account of contextual requirements. So that in so far as the search for the meaning or the search for the analysis of an expression is conceived as the search for a definition, the neglect or misinterpretation of conventions other than ascriptive is inevitable. Perhaps it would be better to say (for I do not wish to legislate about 'meaning' or 'analysis') that logicians have failed to notice that problems of use are wider than problems of analysis and meaning. (2) The influence of the preoccupation with mathematics and formal logic is most clearly seen (to take no more recent examples) in the cases of Leibniz and Russell. The constructor of calculuses, not concerned or required to make factual statements, approaches applied logic with a prejudice. It is natural that he should assume that the types of convention with whose adequacy in one field he is familiar should be really adequate, if only one could see how, in a quite different field—that of statements of fact. Thus we have Leibniz striving desperately to make the uniqueness of unique references a matter of logic in the narrow sense, and Russell striving desperately to do the same thing, in a different way, both for the implication of uniqueness and for that of existence.

It should be clear that the distinction I am trying to draw is primarily one between different roles or parts that expressions may play in language, and not primarily one between different groups of expressions; for some expressions may appear in either role. Some of the kinds of words I shall speak of have predominantly, if not exclusively, a referring role. This is most obviously true of pronouns and ordinary proper names. Some can occur as wholes or parts of expressions which have a predominantly referring use, and as wholes or parts of expressions which have a predominantly ascriptive or classificatory use. The obvious cases are common nouns; or common nouns preceded by adjectives, including participial adjectives; or, less obviously, adjectives or participial adjectives alone. Expressions capable of having a referring use also differ from one another in at least the three following, not mutually independent, ways:

(1) They differ in the extent to which the reference they are used to make is dependent on the context of their utterance. Words like 'I' and 'it' stand at one end of this scale—the end of maximum dependence—and phrases like 'the author of Waverley' and 'the eighteenth king of France' at the other.

(2) They differ in the degree of 'descriptive meaning' they possess: by 'descriptive meaning' I intend 'conventional limitation, in application, to things of a certain general kind, or possessing certain general characteristics'. At one end of this scale stand the proper names we most commonly use in ordinary discourse; men, dogs and motor-bicycles may be called 'Horace'. The pure name has no descriptive meaning (except such as it may acquire *as a result of* some one of its uses as a name). A word like 'he' has minimal descriptive meaning, but has some. Substantival phrases like 'the round table' have the maximum descriptive meaning. An interesting intermediate position is occupied by 'impure' proper names like 'the Round Table'—substantival phrases which have grown capital letters.

(3) Finally, they may be divided into the following two classes: (i) those of which the correct referring use is regulated by some *general* referring-cum-ascriptive conventions. To this class belong both pronouns, which have the least descriptive meaning, and substantival phrases which have the most; (ii) those of which the correct referring use is regulated by no general conventions, either of the contextual or the ascriptive kind, but by conventions which are *ad hoc* for each particular use (though not for each particular utterance). Roughly speaking, the most familiar kind of proper names belong to this class. Ignorance of a man's name is not ignorance of the language. This is why we do not speak of the meaning of proper names. (But it won't do to say they are meaningless.) Again an intermediate position is occupied by such phrases as 'The Old Pretender'. Only an old pretender may be so referred to; but to know which old pretender is not to know a general, but an *ad hoc*, convention.

In the case of phrases of the form 'the so-and-so' used referringly, the use of 'the' together with the position of the phrase in the sentence (i.e. at the beginning, or following a transitive verb or preposition) acts as a signal *that* a unique reference is being made; and the following noun, or noun and adjective, together with the context of utterance, shows *what* unique reference is being made. In general the functional difference between common nouns and adjectives is that the former are naturally and commonly used referringly, while the latter are not commonly, or so naturally, used in this way, except as qualifying nouns; though they can be, and are, so used alone. And of course this functional difference is not independent of the descriptive force peculiar to each word. In general we should expect the descriptive force of nouns to be such that they are more efficient tools for the job of showing what unique reference is intended when such a reference is signalized; and we should also expect the descriptive force of the words we naturally and commonly use to make unique reference to mirror our interest in the salient, relatively permanent and behavioural characteristics of things. These two expectations are not independent of one another; and, if we look at the difference between the commoner sort of common nouns and the commoner sort of adjectives, we find them both fulfilled. These are differences of the kind that Locke quaintly reports, when he speaks of our ideas of substances being *collections* of simple ideas; when he says that 'powers make up a great part of our ideas of substances'; and when he goes on to contrast the identity of real and nominal essence in the case of simple ideas with their lack of identity and the shiftingness of the nominal essence in the case of substances. 'Substance' itself is the troublesome tribute Locke pays to his dim awareness of the difference in predominant linguistic function that lingered even when the noun had been expanded into a more or less indefinite string of adjectives. Russell repeats Locke's mistake with a difference when, admitting the inference from syntax to reality to the extent of feeling that he can get rid of this metaphysical unknown only if he can purify language of the referring function altogether, he draws up his programme for 'abolishing particulars'; a programme, in fact, for abolishing the distinction of logical use which I am here at pains to emphasize.

The contextual requirement for the referring use of pronouns may be stated with the greatest precision in some cases (e.g. 'I' and 'you') and only with the greatest vagueness in others ('it' and 'this'). I propose to say nothing further about pronouns, except to point to an additional symptom of the failure to recognize the uniquely referring use for what it is; the fact, namely, that certain logicians have actually sought to elucidate the nature of the variable by offering such *sentences* as 'he is sick', 'it is green', as examples of something in ordinary speech like a *sentential function*. Now of course it is true that the word 'he' may be used on different occasions to refer to different people or different animals: so may the word 'John' and the phrase 'the cat'. What deters such logicians from treating these two expressions as quasi-variables is, in the first case, the lingering superstition that a name is logically tied to a single individual, and, in the second case, the descriptive meaning of the word 'cat'. But 'he', which has a wide range of applications and minimal descriptive force, only acquires a use as a referring word. It is this fact, together with the failure to accord to expressions used referringly the place in logic which belongs to them (the place held open for the mythical logically proper name), that accounts for the misleading attempt to elucidate the nature of the variable by reference to such words as 'he', 'she', 'it'.

Of ordinary proper names it is sometimes said that they are essentially words each of which is used to refer to just one individual. This is obviously false. Many ordinary personal names—names par excellence—are correctly used to refer to numbers of people. An ordinary personal name, is, roughly, a word, used referringly, of which the use is *not* dictated by any descriptive meaning the word may have, and is *not* prescribed by any such general rule for use as a referring expression (or a part of a referring expression) as we find in the case of such words as 'I', 'this' and 'the', but is governed by *ad hoc* conventions for each particular set of applications of the word to a given person. The important point is that the correctness of such applications does not follow from any *general* rule or convention for the use of the word as such. (The limit of absurdity and obvious circularity is reached in the attempt to treat names as disguised description in Russell's sense; for what is in the special sense implied, but not entailed, by my now referring to someone by name is simply the existence of someone, *now being referred to*, who is *conventionally referred to* by that name.) Even this feature of names, however, is only a symptom of the purpose for which they are employed. At present our choice of names is partly arbitrary, partly dependent on legal and social observances. It would be perfectly possible to have a thorough-going *system* of names, based e.g. on dates of birth, or on a minute classification of physiological and anatomical differences. But the success of any such system would depend entirely on the convenience of the resulting name-allotments for the purpose of making unique references; and this would depend on the multiplicity of the classifications used and the degree to which they cut haphazard across normal social groupings. Given a sufficient degree of both, the selectivity supplied by context would do the rest; just as is the case with our present naming habits. Had we such a system, we could use name-words descriptively (as we do at present, to a limited extent and in a different way, with some famous names) as well as referringly. But it is by criteria derived from consideration of the requirements of the referring task that we should assess the adequacy of any system of naming. From the naming point of view, no kind of classification would be better or worse than any other simply because of the kind of classification—natal or anatomical—that it was.

I have already mentioned the class of quasi-names, of substantival phrases which grow capital letters, and of which such phrases as 'the

Glorious Revolution', 'the Great War', 'the Annunciation', 'the Round Table' are examples. While the descriptive meaning of the words which follow the definite article is still relevant to their referring role, the capital letters are a sign of that extra-logical selectivity in their referring use, which is characteristic of pure names. Such phrases are found in print or in writing when one member of some class of events or things is of quite outstanding interest in a certain society. These phrases are embryonic names. A phrase may, for obvious reasons, pass into, and out of, this class (e.g. 'the Great War').

V

I want to conclude by considering, all too briefly, three further problems about referring uses:

(*a*) *Indefinite references*. Not all referring uses of singular expressions forestall the question 'What (who, which one) are you talking about?' There are some which either invite this question, or disclaim the intention or ability to answer it. Examples are such sentence-beginnings as 'A man told me that . . .' The orthodox (Russellian) doctrine is that such sentences are existential, but not uniquely existential. This seems wrong in several ways. It is ludicrous to suggest that part of what is asserted is that the class of men or persons is not empty. Certainly this is *implied* in the by now familiar sense of implication; but the implication is also as much an implication of the *uniqueness* of the particular object of reference as when I begin a sentence with such a phrase as 'the table'. The difference between the use of the definite and indefinite articles is, very roughly, as follows. We use 'the' either when a previous reference has been made, and when 'the' signalizes that the same reference is being made; or when, in the absence of a previous indefinite reference, the context (including the hearer's assumed knowledge) is expected to enable the hearer to tell what reference is being made. We use 'a' either when these conditions are not fulfilled of when, although a definite reference could be made, we wish to keep dark the identity of the individual to whom, or to which, we are referring. This is the *arch* use of such a phrase as 'a certain person' or 'someone'; where it could be expanded, not into 'someone, but you wouldn't (or I don't) know who' but into 'someone, but I'm not telling you who'.

(*b*) *Identification statements*. By this label I intend statements like the following:

(i*a*) That is the man who swam the channel twice on one day.
(ii*a*) Napoleon was the man who ordered the execution of the Duc D'Enghien.

The puzzle about these statements is that their grammatical predicates do not seem to be used in a straightforwardly ascriptive way as are the grammatical predicates of the statements:

(i*b*) That man swam the channel twice in one day.
(ii*b*) Napoleon ordered the execution of the Duc D'Enghien.

But if, in order to avoid blurring the difference between (i*a*) and (i*b*) and (ii*a*) and (ii*b*), one says that the phrases which form the grammatical complements of (i*a*) and (ii*a*) are being used referringly, one becomes puzzled about what is being said in these sentences. We seem then to be referring to the same person twice over and either saying nothing about him and thus making no statement, or identifying him with himself and thus producing a trivial identity.

The bogey of triviality can be dismissed. This only arises for those who think of the object referred to by the use of an expression as its meaning, and thus think of the subject and complement of these sentences as meaning the same because they could be used to refer to the same person.

I think the differences between sentences in the (*a*) group and sentences in the (*b*) group can best be understood by considering the differences between the circumstances in which you would say (i*a*) and the circumstances in which you would say (i*b*). You would say (i*a*) instead of (i*b*) if you knew or believed that your hearer knew or believed that *someone* had swum the channel twice in one day. You say (i*a*) when you take your hearer to be in the position of one who can ask: 'Who swam the channel twice in one day?' (And in asking this, he is not saying that anyone did, though his asking it implies—in the relevant sense—that someone did.) Such sentences are like answers to such questions. They are better called 'identification-statements' than 'identities'. Sentence (i*a*) does not assert more or less than sentence (i*b*). It is just that you say (i*a*) to a man whom you take to know certain things that you take to be unknown to the man to whom you say (i*b*).

This is, in the barest essentials, the solution to Russell's puzzle about 'denoting phrases' joining by 'is'; one of the puzzles which he claims for the Theory of Descriptions the merit of solving.

(*c*) *The logic of subjects and predicates*. Much of what I have said of the uniquely referring use of expressions can be extended, with suitable modifications, to the non-uniquely referring use of expressions; i.e. to some uses of expressions consisting of 'the', 'all the', 'all', 'some', 'some of the', etc. followed by a noun, qualified or unqualified, in the *plural*; to some uses of 'they', 'them', 'those', 'these'; and to conjunctions of names. Expressions of the first kind have a special interest. Roughly speaking, orthodox modern criticism, inspired by mathematical logic, of such traditional doctrines as that of the Square of Opposition and of some of the forms of the syllogism traditionally recognized as valid, rests on the familiar failure to recognize the special sense in which existential assertions may be implied by the referring use of expressions. The universal propositions of the fourfold schedule, it is said, must *either* be given a negatively existential interpretation (e.g. for A, 'there are no Xs which are not Ys') *or* they must be interpreted as conjunctions of negatively and positively existential statements of, e.g., the form (for A) 'there are no Xs which are not Ys, and there are Xs'. The I and O forms are normally given a positively existential interpretation. It is then seen that, whichever of the above alternatives is selected, some of the traditional laws have to be abandoned. The dilemma, however, is a bogus one. If we interpret the propositions of the schedule as neither positively, nor negatively, nor positively *and* negatively, existential, but as sentences such that *the question of whether they are being used to make true or false assertions does not arise except when the existential condition is fulfilled for the subject term*, then all the traditional laws hold good together. And this interpretation is far closer to the most common uses of expressions beginning with 'all' and 'some' than is any Russellian alternative. For these expressions are most commonly used in the referring way. A literal-minded and childless man asked whether all his children are asleep will certainly not answer 'Yes' on the ground that he has none; but nor will he answer 'No' on this ground. Since he has no children, the question does not arise. To say this is not to say that I may not use the sentence, 'All my children are asleep', with the intention of letting someone know that I have children, or of deceiving him into thinking that I have. Nor is it any weakening of my thesis to concede that singular phrases of the form 'the so-and-so' may sometimes be used with a similar purpose. Neither Aristotelian nor Russellian rules give the exact logic of any expression of ordinary language; for ordinary language has no exact logic.

V

MIND AND VERBAL DISPOSITIONS

W. V. QUINE

Descartes supposed that man is the only animal endowed with mind; the others are automata. It is held further, and more widely and on better evidence, that man is the only animal endowed with language. Now if man is unique in enjoying these two gifts, it is no coincidence. One may argue that no mindless creature could cope with so intricate a device as language. Or one may argue conversely that no appreciable mental activity is conceivable without linguistic aids.

Most thought simply *is* speech, according to the pioneer behaviourist John B. Watson: silent, repressed, incipient speech. Not all thought is that. A geometer or an engineer may think by means also of little incipient tugs of the muscles that are used in drawing curves of twirling cogwheels. Still, the muscles that play by far the major role, according to Watson's muscular theory of meditation, are the muscles used in making speeches.

Conversely, there is an age-old and persistent tendency to try to explain and analyse the physical phenomenon of speech by appealing to mind, mental activity, and mental entities: by appealing to thoughts, ideas, meanings. Language, we are told, serves to convey ideas. We learn language from our elders by learning to associate the words with the same ideas with which our elders have learned to associate them. Thus it is, we may be told, that approximate uniformity of association of words with ideas is achieved and maintained throughout the community.

Such an account would of course be extravagantly perverse. Thus consider the case where we teach the infant a word by reinforcing his random babbling on some appropriate occasion. His chance utterance bears a chance resemblance to a word appropriate to the occasion, and we reward him. The occasion must be some object or some stimulus source that we as well as the child are in a position to notice. Furthermore, we must be in a position to observe that the child is in a position to notice it. Only thus would there be any purpose in our rewarding his chance utterance. In so doing we encourage the child to repeat the word

From Samuel Guttenplan (ed.), *Mind and Language* (Oxford: Clarendon Press, 1975).

on future similar occasions. But are we causing him to associate the word with the same *idea* that we adults associate it with? Do we adults all associate it with the same idea ourselves, for that matter? And what would that mean?

The moral of this is that the fixed points are just the shared stimulus and the word; the ideas in between are as may be and may vary as they please, so long as the external stimulus in question stays paired up with the word in question for all concerned. The point is well dramatized by the familiar fantasy of complementary colour perception. Who knows but that I see things in colours opposite to those in which you see the things? For communication it is a matter of indifference.

I believe in the affinity of mind and language, but I want to keep the relation right side up. Watson's theory of thought, however inadequate, has matters right side up. A theory of mind can gain clarity and substance, I think, from a better understanding of the workings of language, whereas little understanding of the workings of language is to be hoped for in mentalistic terms.

I shall say a little about how it is that people feel drawn to a mentalistic account of language, despite the conspicuous fact that language is a social enterprise which is keyed to intersubjectively observable objects in the external world. Also I shall speculate on how we might hope to get on with a properly physicalistic account of language. First I must talk a little more about learning.

I mentioned one primitive way of learning a word: through reinforcement of random babbling. Another way, somewhat the reverse, is imitation. In the case of babbling it was the adult that witnessed what was confronting the child when the child chanced to babble the appropriate word. In the case of imitation it is the child, conversely, that witnesses what is confronting the adult when the adult volunteers the word. The child then volunteers the word when similarly confronted, and thereupon the adult proceeds to reinforce the child's behaviour just as in the case of babbling. The imitation method is more sophisticated than the babbling method. It can still be explained, indirectly, in terms of stimulus and reinforced response; but I won't pause over it.

What we do need to notice is that all language learning at this primitive level is directed only to the learning of what may be called observation terms, or, more properly, observation sentences. The child learns to assent to the query 'red?' in the conspicuous presence of red things. Also he masters the trick of getting the object by uttering the word; 'red' is a poor example here, but 'ball' and 'milk' and 'Mama' are clear cases. Also he masters the word in a passive way, responding in some distinctive fashion on hearing it. He may respond by turning to face the object, or by fetching it.

Now the observation term, or observation sentence, is a ground on which John the rational animal and Fido the automaton can meet and to some degree communicate. The dog learns observation sentences in his passive way. He learns to respond to them by salivating, by running to the kitchen, by turning to face the object, or by fetching it.

Already at this lowly level of observation sentences the small child differs from the dog, it may seem, in that he learns the sentences also actively: he utters them. This still is not a clear contrast. Dogs learn to ask for things, in their inarticulate way. Let us not arrogate to rationality what may be merely superior agility of lips and tongue and larynx. Premack and his chimpanzee have circumvented these muscular obstacles by resorting to plastic symbols, which they push around on a board. Premack succeeds in teaching his chimpanzee to volunteer observation sentences appropriately and to play a passable game of query and assent.

A contrast that has long been remarked, between human language and animal signals, is the combinatorial productivity of language: man's ability to compose new and unprecedented sentences from old materials, and to respond appropriately to such new creations. But Premack reports that his chimpanzee even passes this test, within modest limits. It would thus appear that combinatorial productivity in language affords no sharp line between man and beast. Man may plume himself on having been the first to develop a combinatorially productive language, but the ability to learn it may be more widespread.

Combinatorial productivity, however, is not the only trait that has seemed to distinguish mind-governed discourse from the performance of trained animals. A major factor is the unpredictable spontaneity of speech. Animal drives are still at work behind the torrent of human speech, but they are seldom clearly to be traced. Even if in our verbal output we differ from Premack's chimpanzee only in degree and not in kind, still it is this overwhelming difference of degree that invites the mentalistic accounts of verbal behaviour. The torrent of words is seen as a manifestation of the speaker's inner life beyond animal drives. Nowadays one is apt to resort thus to a mentalistic semantics not so much because one sees an ontological gulf between man and the apes, as because one despairs of adhering to the standards of natural science in coping with the complexities of intelligent discourse.

The central notion in mentalistic semantics is an unanalysed notion of meaning. It figures mainly in two contexts: where we speak of knowing the meaning of an expression, and where we speak of sameness of meaning. We say we know the meaning of an expression when we are able to produce a clearer or more familiar expression having the same meaning. We ask the meaning of an expression when what we want is a clearer or more familiar expression having the same meaning.

I said to my small son, 'Eighty-two. You know what I mean?' He said, 'No.' Then I said to my small daughter, 'Ottantadue. You know what I mean?' She said, 'Yes. Eighty-two.' I said, 'See, Margaret understands Italian better than Douglas understands English.'

Our ways of talking of meaning are thus misleading. To understand an expression is, one would say, to know the meaning; and to know the meaning is, one would say, to be able to give the meaning. Yet Douglas could rightly claim to *understand* the expression 'eighty-two', despite answering 'No' to 'You know what I mean?' He answered 'No' because he was unable to *give* the meaning; and he was unable to give the meaning because what we call giving the meaning consists really in the asymmetrical operation of producing an equivalent expression that is clearer. Margaret was ready with a clearer equivalent of 'ottantadue', but Douglas was at a loss for a *still* clearer equivalent of 'eighty-two'. In another context he might have ventured, 'Yes, you mean the temperature is eighty-two.'

People persist in talking thus of knowing the meaning, and of giving the meaning, and of sameness of meaning, where they could omit mention of meaning and merely talk of understanding an expression, or talk of the equivalence of expressions and the paraphrasing of expressions. They do so because the notion of meaning is felt somehow to *explain* the understanding and equivalence of expressions. We understand expressions by knowing or grasping their meanings; and one expression serves as a translation or paraphrase of another because they mean the same. It is of course spurious explanation, mentalistic explanation at its worst. The paradoxical little confusion between understanding 'eighty-two' and knowing or giving its meaning is always symptomatic of awkward concept-building; but where the real threat lies, in talking of meaning, is in the illusion of explanation.

In all we may distinguish three levels of purported explanation, three degrees of depth: the mental, the behavioural, and the physiological. The mental is the most superficial of these, scarcely deserving the name of explanation. The physiological is the deepest and most ambitious, and it is the place for causal explanations. The behavioural level, in between, is what we must settle for in our descriptions of language, in our formulations of language rules, and in our explications of semantical terms. It is here, if anywhere, that we must give our account of the understanding of an expression, and our account of the equivalence that holds between an expression and its translation or paraphrase. These things need to be explained, if at all, in behavioural terms: in terms of dispositions to overt gross behaviour.

Take understanding. Part of the understanding of a word consists in the ability to use it properly in all manner of admissible contexts. Part consists in reacting properly to all such uses. So there is a good deal here to sort out and organize. We must divide and define. To begin with we can set aside the complication of the myriad sentential contexts of a word, by beginning rather with sentences as wholes: with complete little isolated speeches, consisting perhaps of a single word and perhaps of more.

Bewildering variety confronts us even so. One and the same little sentence may be uttered for various purposes: to warn, to remind, to obtain possession, to gain confirmation, to gain admiration, or to give pleasure by pointing something out. The occasions for uttering one and the same sentence are so various that we can seldom predict when a sentence will be uttered or which one it will be. An unpromising setting, this, in which to explore and exploit verbal dispositions. Somehow we must further divide; we must find some significant central strand to extract from the tangle.

Truth will do nicely. Some sentences, of course, do not have truth values: thus questions and imperatives. Those that do may still be uttered for a variety of reasons unconnected with instruction; I just now enumerated some. But, among these sentences, truth is a great leveller, enabling us to postpone consideration of all those troublesome excrescences. Here, then, is an adjusted standard of understanding: a man understands a sentence in so far as he knows its truth conditions. This kind of understanding stops short of humour, irony, innuendo, and other literary values, but it goes a long way. In particular it is all we can ask of an understanding of the language of science.

We are interested not only in explaining what it is for someone else to understand a sentence, but also in setting a standard for ourselves, as when we try to penetrate a new language and to understand its sentences, or try to teach the language. Our standard, still, is this: give the truth conditions. Hence Davidson's plan for a semantics in the style of Tarski's truth definition.

But when I define the understanding of a sentence as knowledge of its truth conditions I am certainly not offering a definition to rest with; my term 'knowledge' is as poor a resting-point as the term 'understanding' itself.

We were supposed to be getting things down to terms of dispositions to behaviour. In what behavioural disposition then does a man's knowledge of the truth conditions of the sentence 'This is red' consist? Not, certainly, in a disposition to affirm the sentence on every occasion of observing a red object, and to deny it on all other occasions; it is the disposition to assent or dissent when asked in the presence or absence of red. Query and assent, query and dissent—here is the solvent that reduces understanding to verbal disposition. Without this device there would be no hope of handing language down the generations, nor any hope of breaking into newly discovered languages. It is primarily by querying sentences for assent and dissent that we tap the reservoirs of verbal disposition.

This approach applies primarily to terms, or occasion sentences, rather than to standing sentences. For the disposition to assent to or dissent from the sentence 'This is red' is marked by a correlation between assent and the presence of red, and between dissent and the absence of red, on occasions where the sentence is queried. A standing sentence, whose truth value remains fixed over long periods, offers no significant correlation of the kind. Where the method of queried assent and dissent is at its best, indeed, is in application to occasion sentences of the special sort that I have called observation sentences; for the occasions that make the sentence true are going to have to be intersubjectively recognizable if we are to be able to *tell* whether the speaker has the disposition in question. Even in these cases, of course, we remain at the mercy of the speaker's veracity: we assume when querying him that his assents and dissents are sincere. Happily we live in a moral climate where this assumption generally holds up; language could not flourish otherwise.

Standing sentences can be queried too, but the stimulating situation at the time of querying them will usually have no bearing on the verdict; and for this reason we cannot identify the understanding of a standing sentence, even approximately, with a disposition to assent or dissent when queried on particular occasions. I do not know how, in general, in terms of behavioural dispositions, to approximate to the notion of understanding at all, when the sentences understood are standing sentences. Perhaps it cannot be done, taking standing sentences one by one.

Once in a while we get a hint of a specifically relevant disposition, still, when we find someone reversing his verdict on a standing sentence in the face of some observation. But with all conceivable luck we cannot hope to correlate standing sentences generally with observations, because the sentences one by one simply do not have their own separable empirical implications. A multiplicity of standing sentences will interlock, rather, as a theory; and an observation in conflict with that theory may be accommodated by revoking one or other of the sentences—no one sentence in particular.

One sees how a semanticist might despair and seek shelter in the jungle of mentalistic semantics. But there are other courses. Perhaps the very notion of understanding, as applied to single standing sentences, simply cannot be explicated in terms of behavioural dispositions. Perhaps therefore it is simply an untenable notion, notwithstanding our intuitive predilections. It stands to reason that a proper semantical analysis of standing sentences, in terms of behavioural dispositions, will be primarily occupied with the interrelations of sentences rather than with standing sentences one by one.

I mentioned two central semantical notions which, in mentalistic semantics, are obscured by talk of meaning. One was the notion of understanding an expression, and the other was the relation of equivalence between an expression and its paraphrase. Afterwards I considered what might be done about understanding. Now what about the other notion, the equivalence relation? Much of what I have said about understanding applies in parallel fashion to equivalence. Here, as there, we can conveniently organize our work by looking first to sentences as wholes, seeking an equivalence concept for them. Here, as there, we can usefully narrow our problem by focusing on truth conditions and so exploiting a method of query and assent. And here, of course, as there, the sentences that prove reasonably amenable are the occasion sentences, especially the observation sentences. What relates such a sentence to its equivalent is simply a coinciding of dispositions: we are disposed to assent to both sentences in the same circumstances.

Moreover, in a behavioural account of equivalence, just as in a behavioural account of understanding, we encounter difficulty when we move to standing sentences. Since a man is apt to assent to a standing sentence, if asked, in all sorts of circumstances or in none, the coinciding of dispositions to assent to two standing sentences gives no basis for equating them.

I am persuaded, indeed, that a satisfactory equivalence concept is impossible for standing sentences. My view of this matter can be conveyed most clearly if we consider translation between two languages. I am persuaded that alternative manuals of translation can exist, incompatible with each other, and both of them conforming fully to the dispositions to behaviour on the part of the speakers of the two languages. The two manuals would agree on observation sentences but conflict in some of the standing sentences. Each manual, being a manual of translation, purports to specify the equivalence relation between sentences and their translations, and neither manual is right to the exclusion of the other.

This indeterminacy of translation is unsuspected in mentalistic semantics, because of the facile talk of meaning. Sentences have meanings, and a translation is right if it has the same meaning. Mentalistic semantics requires that one of two conflicting manuals of translation be wrong, though it conforms to every speaker's dispositions. Mentalistic semantics thus sets a false goal, which, even though vague and ill defined, tends to obstruct other lines of thought.

Of course, translation must go on. Indeterminacy means that there is more than one way; we can still proceed to develop one of them, as good as any. And, in a more theoretical mood, we must still consider what counts as evidence for *an* acceptable translation relation, even if the relation is not unique. The evidence will be behavioural still, of course, even though the relation is no simple coinciding of behavioural dispositions, as it was in the case of the equivalence of observation sentences. We have to examine relations of interdependence between verbal dispositions: systematic interdependences between dispositions to assent to standing sentences and dispositions to assent in certain circumstances to observation sentences. Here again, in the problem of equivalence as in the problem of understanding, it would seem that genetic semantics offers a likely approach. But we must expect no simple picture, no easy answers. For it is a question again of the relations of standing sentences to observation sentences, and hence nothing less than the relation of scientific theory to scientific evidence.

Let us then recognize that the semantical study of language is worth pursuing with all the scruples of the natural scientist. We must study language as a system of dispositions to verbal behaviour, and not just surface listlessly to the Sargasso Sea of mentalism.

It has been objected that when I talk of query and assent I am not really escaping mentalism after all, because assent itself has a mental component. It is objected that assent is no mere mindless parroting of an arbitrary syllable; utterance of the syllable counts as assent only if there is the appropriate mental act behind it. Very well, let us adopt the term *surface assent* for the utterance or gesture itself. My behavioural approach does indeed permit me, then, only to appeal to surface assent; assent as I talk of it must be understood as surface assent. This behavioural notion has its powers, however, and must not be underrated. For the syllable or gesture of assent in a community is not identified at random, after all; it is itself singled out, in turn, by behavioural criteria. One partial criterion of what to count as a sign of assent is that a speaker is disposed to produce that sign whenever a sentence is queried in circumstances in which he would be disposed to volunteer the sentence himself. Even surface assent, thus, is not just the parroting of any arbitrary syllable. Granted, some cases of surface assent are insincere, but happily they are rare enough, as I have already remarked, to permit the field linguist still to find laws and translations on the strength of statistical trends.

I have been inveighing against mentalistic semantics and urging in its place the study of dispositions to behaviour. This move could be represented alternatively and more picturesquely as a matter not so much of substitution as of identification: let us *construe* mind as a system of disposition to behaviour. This version somewhat recalls Gilbert Ryle and Wilfrid Sellars, who have urged a generally dispositional philosophy of mind. Some small further encouragement for it may be seen in the fact that even our most ordinary and characteristic mentalistic idioms already take almost the form of attributions of verbal dispositions. These are the idioms of propositional attitude: '*x* believes that *p*', '*x* wishes that *p*', '*x* expects that *p*', and so on. They all follow the broad pattern of indirect quotation, '*x* says that *p*', as if to attribute to *x* the disposition to utter the sentence '*p*' in some mood. Thus *x* believes that *p* if, approximately, he will affirm *p*; he wishes or regrets that *p* if, approximately, he will exclaim 'Oh that *p*!' or 'Alas, *p*!'

I am offering no proper analysis of the propositional attitudes. People do not volunteer all their beliefs in affirmations. A better criterion of belief is the disposition to assent if asked; and this still leaves no room for questioning sincerity. Also there is the problem of allowable latitude of translation or paraphrase, when the '*p*' clause in '*x* believes that *p*' contains words alien to *x*'s actual vocabulary. This question of allowable latitude of course arises acutely in indirect discourse itself, '*x* said that *p*', and it plagues all the idioms of propositional attitude. And finally there are quandaries over the referential opacity of the idioms of propositional attitude: quandaries having to do with the substitutivity of identity, and with quantifying into opaque contexts. All in all, the propositional attitudes are in a bad way. These are the idioms most stubbornly at variance with scientific patterns. Consequently I find it particularly striking that these, of all idioms, already describe mental states in a way that hints at dispositions to verbal behaviour. A philosophy of mind as verbal disposition is after all not so very alien to deep-rooted popular attitudes.

I spoke of three levels of purported explanation: the mental, the behavioural, and the physiological. We have just now been contemplating the second, the behavioural. Now the relation of this level to the third and deepest, the physiological, begins to be evident when we examine the notion of a *disposition* to behaviour and consider what we mean by a disposition.

A disposition is in my view simply a physical trait, a configuration or mechanism. It can be a disjunctive physical trait, since like effects can come of unlike mechanisms. What makes it a disposition is no significant character of its own, but only the style in which we happen to specify it. Thus take the classical example, solubility in water. This is a physical trait that can be specified, with varying degrees of thoroughness, in various ways. It can be described quite fully, I gather, in terms of the relative positions of small particles. It can also be described, less fully, by citing a simple test: put an object in water and see if it dissolves. Instructions for this convenient test are compactly encoded, as it happens, in the adjective 'soluble' itself, with its verb stem 'solu-' and its dispositional ending '-ble'. The adjective 'soluble' is a disposition word, and this is an important classification of words; but the physical traits themselves do not divide into dispositions and others, any more significantly than mankind divides into passers-by and others. The term 'disposition' has its significant application rather as a preface, each time, to an actual singling out of some physical trait; thus we may significantly specify some physical trait as the disposition to behave thus and so in such-and-such circumstances. It is this that is accomplished also, and more laconically, by dispositional adjectives such as 'soluble', 'fragile', 'docile', 'portable'. The dispositional way of specifying physical traits is as frequent and as useful as it is because we are so often not prepared, as we now happen to be in the case of solubility, to specify the intended physical trait in other than the dispositional style.

The dispositional way of specifying physical states and traits is indeed pretty generally *the* way of specifying them, except at high levels of scientific theory. The explicit dispositional idiom does not always appear, either as the word 'disposition' or as a suffix '-ble' or '-ile'; commonly the dispositional force is only implicit. Hardness, for instance, is the disposition to resist if pressed, or to scratch. Redness, said of a body, is the disposition to blush in white light. Hardness and redness come finally, like solubility, to be explained in terms of minute structure, but our first access to these physical traits is dispositional. In fact the same may be said of the very notion of body itself; for a body comes to be known, as Kant remarked, by its disposition to present a repeatable sequence of views as we walk around it or revisit it. True to form, even this disposition qualifies as a physical mechanism: *body*. Like the other physical mechanisms, this one also comes in the fulness of time to be explained in terms of small particles.

John Stuart Mill's characterization of a body as 'a permanent possibility of sensation' was meant in an idealistic spirit, as a reduction of matter to sensory disposition. Thanks to symmetry, however, the identity admits also of a materialistic inversion: corporeality, like solubility, is an objective physical arrangement of particles, but known first in dispositional terms.

Dispositions to behaviour, then, are physiological states or traits or mechanisms. In citing them dispositionally we are singling them out by behavioural symptoms, behavioural tests. Usually we are in no position to detail them in physiological terms, but in this there is no anomaly; we also commonly specify ailments *per accidens*, citing gross signs and symptoms and knowing no physiological details.

We now see the relation of the second level of explanation, the behavioural, to the third and deepest level, the physiological. At the second level we treat of dispositions to behaviour, and these dispositions are indeed physiological states, but we identify them only by their behavioural manifestations. The deepest explanation, the physiological, would analyse these dispositions in explicit terms of nerve impulses and other anatomically and chemically identified organic processes.

Our three levels thus are levels of reduction: mind consists in dispositions to behaviour, and these are physiological states. We recall that John B. Watson did not claim that quite *all* thought was incipient speech; it was all incipient twitching of muscles, and *mostly* of speech muscles. Just so, I would not identify mind quite wholly with verbal disposition; with Ryle and Sellars I would identify it with behavioural disposition, and *mostly* verbal. And then, having construed behavioural dispositions in turn as physiological states, I end up with the so-called identity theory of mind: mental states are states of the body.

However, a word of caution is in order regarding the so-called identity theory. How does it differ from a repudiation theory? Let us think for a moment about an analogous question elsewhere, concerning the definitions of natural number in set theory. We may say that numbers are defined as sets in Frege's way, or in Zermelo's way, or in von Neumann's way, these ways all being good but incompatible. Or we may say that numbers may be repudiated, dispensed with ; that we can get along with just sets instead, in any of various ways—Frege's way, Zermelo's way, von Neumann's way. This repudiation version has the advantage that we no longer seem called upon to adjudicate between three identifications of the natural numbers, the three being incompatible and yet all somehow correct.

Correspondingly, instead of saying that mental states are identical with physiological ones, we could repudiate them; we could claim that they can be dispensed with, in all our theorizing, in favour of physiological states, these being specified usually not in actual physiological terms but in the idiom of behavioural dispositions. This repudiation version has a certain advantage, though a different one from what we noted in the case of number. Its advantage here is that it discourages a possible abuse of the identity theory. For, product though the identity theory is of hard-headed materialism, we must beware of its sedative use to relieve intellectual discomfort. We can imagine someone appealing to the identity theory to excuse his own free and uncritical recourse to mentalistic semantics. We can imagine him pleading that it is after all just a matter of physiology, even if no one knows quite how. This would be a sad irony indeed, and the repudiation theory has the virtue, over the identity theory, of precluding it.

Until we can aspire to actual physiological explanation of linguistic activity in physiological terms, the level at which to work is the middle one; that of dispositions to overt behaviour. Its virtue is not that it affords causal explanations but that it is less likely than the mentalistic level to engender an illusion of being more explanatory than it is. The easy familiarity of mentalistic talk is not to be trusted.

Still, among the dispositions to behaviour, some are more explanatory than others. The ones that we should favour, in explanations, are the ones whose physiological mechanisms seem likeliest to be detected in the foreseeable future. To cite a behavioural disposition is to posit an unexplained neural mechanism, and such posits should be made in the hope of their submitting some day to a physical explanation.

VI

TRUTH AND MEANING

DONALD DAVIDSON

It is conceded by most philosophers of language, and recently by some linguists, that a satisfactory theory of meaning must give an account of how the meanings of sentences depend upon the meanings of words. Unless such an account could be supplied for a particular language, it is argued, there would be no explaining the fact that we can learn the language: no explaining the fact that, on mastering a finite vocabulary and a finitely stated set of rules, we are prepared to produce and to understand any of a potential infinitude of sentences. I do not dispute these vague claims, in which I sense more than a kernel of truth.[1] Instead I want to ask what it is for a theory to give an account of the kind adumbrated.

One proposal is to begin by assigning some entity as meaning to each word (or other significant syntactical feature) of the sentence; thus we might assign Theaetetus to 'Theaetetus' and the property of flying to 'flies' in the sentence 'Theaetetus flies'. The problem then arises how the meaning of the sentence is generated from these meanings. Viewing concatenation as a significant piece of syntax, we may assign to it the relation of participating in or instantiating; however, it is obvious that we have here the start of an infinite regress. Frege sought to avoid the regress by saying that the entities corresponding to predicates (for example) are 'unsaturated' or 'incomplete' in contrast to the entities that correspond to names, but this doctrine seems to label a difficulty rather than solve it.

The point will emerge if we think for a moment of complex singular terms, to which Frege's theory applies along with sentences. Consider the expression 'the father of Annette'; how does the meaning of the whole depend on the meaning of the parts? The answer would seem to be that the meaning of 'the father of' is such that when this expression is prefixed to a singular term the result refers to the father of the person to

From *Synthese*, 17 (1967), 304–23. Reprinted by permission of Kluwer Academic Publishers. Notes 10, 11, 17, and 20 were added for the reprint in Davidson's *Inquiries into Truth and Interpretation* (Oxford: Clarendon Press, 1984).

[1] See 'Theories of Meaning and Learnable Languages', in D. Davidson, *Inquiries into Truth and Interpretation* (Oxford: Clarendon Press, 1984), 3–15.

whom the singular term refers. What part is played, in this account, by the unsaturated or incomplete entity for which 'the father of' stands? All we can think to say is that this entity 'yields' or 'gives' the father of x as value when the argument is x, or perhaps that this entity maps people on to their fathers. It may not be clear whether the entity for which 'the father of' is said to stand performs any genuine explanatory function as long as we stick to individual expressions; so think instead of the infinite class of expressions formed by writing 'the father of' zero or more times in front of 'Annette'. It is easy to supply a theory that tells, for an arbitrary one of these singular terms, what it refers to: if the term is 'Annette' it refers to Annette, while if the term is complex, consisting of 'the father of' prefixed to a singular term t, then it refers to the father of the person to whom t refers. It is obvious that no entity corresponding to 'the father of' is, or needs to be, mentioned in stating this theory.

It would be inappropriate to complain that this little theory *uses* the words 'the father of' in giving the reference of expressions containing those words. For the task was to give the meaning of all expressions in a certain infinite set on the basis of the meaning of the parts; it was not in the bargain also to give the meanings of the atomic parts. On the other hand, it is now evident that a satisfactory theory of the meanings of complex expressions may not require entities as meanings of all the parts. It behoves us then to rephrase our demand on a satisfactory theory of meaning so as not to suggest that individual words must have meanings at all, in any sense that transcends the fact that they have a systematic effect on the meanings of the sentences in which they occur. Actually, for the case at hand we can do better still in stating the criterion of success: what we wanted, and what we got, is a theory that entails every sentence of the form 't refers to x' where 't' is replaced by a structural description[2] of a singular term, and 'x' is replaced by that term itself. Further, our theory accomplishes this without appeal to any semantical concepts beyond the basic 'refers to'. Finally, the theory clearly suggests an effective procedure for determining, for any singular term in its universe, what that term refers to.

A theory with such evident merits deserves wider application. The device proposed by Frege to this end has a brilliant simplicity: count predicates as a special case of functional expressions, and sentences as a special case of complex singular terms. Now, however, a difficulty looms if we want to continue in our present (implicit) course of identifying the

[2] A 'structural description' of an expression describes the expression as a concatenation of elements drawn from a fixed finite list (for example of words or letters).

meaning of a singular term with its reference. The difficulty follows upon making two reasonable assumptions: that logically equivalent singular terms have the same reference, and that a singular term does not change its reference if a contained singular term is replaced by another with the same reference. But now suppose that '*R*' and '*S*' abbreviate any two sentences alike in truth value. Then the following four sentences have the same reference:

(1) R
(2) $\hat{x}(x = x \,.\, R) = \hat{x}(x = x)$
(3) $\hat{x}(x = x \,.\, S) = \hat{x}(x = x)$
(4) S

For (1) and (2) are logically equivalent, as are (3) and (4), while (3) differs from (2) only in containing the singular term '$\hat{x}(x = x \,.\, S)$' where (2) contains '$\hat{x}(x = x \,.\, R)$' and these refer to the same thing if S and R are alike in truth value. Hence any two sentences have the same reference if they have the same truth value.[3] And if the meaning of a sentence is what it refers to, all sentences alike in truth value must be synonymous—an intolerable result.

Apparently we must abandon the present approach as leading to a theory of meaning. This is the natural point at which to turn for help to the distinction between meaning and reference. The trouble, we are told, is that questions of reference are, in general, settled by extra-linguistic facts, questions of meaning not, and the facts can conflate the references of expressions that are not synonymous. If we want a theory that gives the meaning (as distinct from reference) of each sentence, we must start with the meaning (as distinct from reference) of the parts.

Up to here we have been following in Frege's footsteps; thanks to him, the path is well known and even well worn. But now, I would like to suggest, we have reached an impasse: the switch from reference to meaning leads to no useful account of how the meanings of sentences depend upon the meanings of the words (or other structural features) that compose them. Ask, for example, for the meaning of 'Theaetetus flies'. A Fregean answer might go something like this: given the meaning of 'Theaetetus' as argument, the meaning of 'flies' yields the meaning of 'Theaetetus flies' as value. The vacuity of this answer is obvious. We wanted to know what the meaning of 'Theaetetus flies' is; it is no

[3] The argument derives from Frege. See A. Church, *Introduction to Mathematical Logic* (Princeton, NJ: Princeton University Press, 1956), 24–5. It is perhaps worth mentioning that the argument does not depend on any particular identification of the entities to which sentences are supposed to refer.

progress to be told that it is the meaning of 'Theaetetus flies'. This much we knew before any theory was in sight. In the bogus account just given, talk of the structure of the sentence and of the meanings of words was idle, for it played no role in producing the given description of the meaning of the sentence.

The contrast here between a real and pretended account will be plainer still if we ask for a theory, analogous to the miniature theory of reference of singular terms just sketched, but different in dealing with meanings in place of references. What analogy demands is a theory that has as consequences all sentences of the form '*s* means *m*' where '*s*' is replaced by a structural description of a sentence and '*m*' is replaced by a singular term that refers to the meaning of that sentence; a theory, moreover, that provides an effective method for arriving at the meaning of an arbitrary sentence structurally described. Clearly some more articulate way of referring to meanings than any we have seen is essential if these criteria are to be met.[4] Meanings as entities, or the related concept of synonymy, allow us to formulate the following rule relating sentences and their parts: sentences are synonymous whose corresponding parts are synonymous ('corresponding' here needs spelling out of course). And meanings as entities may, in theories such as Frege's, do duty, on occasion, as references, thus losing their status as entities distinct from references. Paradoxically, the one thing meanings do not seem to do is oil the wheels of a theory of meaning—at least as long as we require of such a theory that it non-trivially give the meaning of every sentence in the language. My objection to meanings in the theory of meaning is not that they are abstract or that their identity conditions are obscure, but that they have no demonstrated use.

This is the place to scotch another hopeful thought. Suppose we have a satisfactory theory of syntax for our language, consisting of an effective method of telling, for an arbitrary expression, whether or not it is independently meaningful (i.e. a sentence), and assume as usual that this involves viewing each sentence as composed, in allowable ways, out of elements drawn from a fixed finite stock of atomic syntactical elements (roughly, words). The hopeful thought is that syntax, so conceived, will yield semantics when a dictionary giving the meaning of each syntactic

[4] It may be thought that Church, in 'A Formulation of the Logic of Sense and Denotation', in P. Henle, H. M. Kallen, and S. K. Langer (eds.), *Structure, Method and Meaning: Essays in Honour of H. M. Sheffer* (New York: Liberal Arts Press, 1951), 3–24, has given a theory of meaning that makes essential use of meanings as entities. But this is not the case: Church's logics of sense and denotation are interpreted as being about meanings, but they do not mention expressions and so cannot of course be theories of meaning in the sense now under discussion.

atom is added. Hopes will be dashed, however, if semantics is to comprise a theory of meaning in our sense, for knowledge of the structural characteristics that make for meaningfulness in a sentence, plus knowledge of the meanings of the ultimate parts, does not add up to knowledge of what a sentence means. The point is easily illustrated by belief sentences. Their syntax is relatively unproblematic. Yet, adding a dictionary does not touch the standard semantic problem, which is that we cannot account for even as much as the truth conditions of such sentences on the basis of what we know of the meanings of the words in them. The situation is not radically altered by refining the dictionary to indicate which meaning or meanings an ambiguous expression bears in each of its possible contexts; the problem of belief sentences persists after ambiguities are resolved.

The fact that recursive syntax with dictionary added is not necessarily recursive semantics has been obscured in some recent writing on linguistics by the intrusion of semantic criteria into the discussion of purportedly syntactic theories. The matter would boil down to a harmless difference over terminology if the semantic criteria were clear; but they are not. While there is agreement that it is the central task of semantics to give the semantic interpretation (the meaning) of every sentence in the language, nowhere in the linguistic literature will one find, so far as I know, a straightforward account of how a theory performs this task, or how to tell when it has been accomplished. The contrast with syntax is striking. The main job of a modest syntax is to characterize *meaningfulness* (or sentencehood). We may have as much confidence in the correctness of such a characterization as we have in the representativeness of our sample and our ability to say when particular expressions are meaningful (sentences). What clear and analogous task and test exist for semantics?[5]

We decided a while back not to assume that parts of sentences have meanings except in the ontologically neutral sense of making a systematic contribution to the meaning of the sentences in which they occur. Since postulating meanings has netted nothing, let us return to

[5] For a recent statement of the role of semantics in linguistics, see Noam Chomsky, 'Topics in the Theory of Generative Grammar', in T. A. Sebeok (ed.), *Current Trends in Linguistics*, iii (The Hague: Mouton, 1966). In this article, Chomsky (1) emphasizes the central importance of semantics in linguistic theory, (2) argues for the superiority of transformational grammars over phrase-structure grammars largely on the grounds that, although phrase-structure grammars may be adequate to define sentencehood for (at least) some natural languages, they are inadequate as a foundation for semantics, and (3) comments repeatedly on the 'rather primitive state' of the concepts of semantics and remarks that the notion of semantic interpretation 'still resists any deep analysis'.

that insight. One direction in which it points is a certain holistic view of meaning. If sentences depend for their meaning on their structure, and we understand the meaning of each item in the structure only as an abstraction from the totality of sentences in which it features, then we can give the meaning of any sentence (or word) only by giving the meaning of every sentence (and word) in the language. Frege said that only in the context of a sentence does a word have meaning; in the same vein he might have added that only in the context of the language does a sentence (and therefore a word) have meaning.

This degree of holism was already implicit in the suggestion that an adequate theory of meaning must entail *all* sentences of the form '*s* means *m*'. But now, having found no more help in meanings of sentences than in meanings of words, let us ask whether we can get rid of the troublesome singular terms supposed to replace '*m*' and to refer to meanings. In a way, nothing could be easier: just write '*s* means that *p*', and imagine '*p*' replaced by a sentence. Sentences, as we have seen, cannot name meanings, and sentences with 'that' prefixed are not names at all, unless we decide so. It looks as though we are in trouble on another count, however, for it is reasonable to expect that in wrestling with the logic of the apparently non-extensional 'means that' we will encounter problems as hard as, or perhaps identical with, the problems our theory is out to solve.

The only way I know to deal with this difficulty is simple, and radical. Anxiety that we are enmeshed in the intensional springs from using the words 'means that' as filling between description of sentence and sentence, but it may be that the success of our venture depends not on the filling but on what it fills. The theory will have done its work if it provides, for every sentence *s* in the language under study, a matching sentence (to replace '*p*') that, in some way yet to be made clear, 'gives the meaning' of *s*. One obvious candidate for matching sentence is just *s* itself, if the object language is contained in the metalanguage; otherwise a translation of *s* in the metalanguage. As a final bold step, let us try treating the position occupied by '*p*' extensionally: to implement this, sweep away the obscure 'means that', provide the sentence that replaces '*p*' with a proper sentential connective, and supply the description that replaces '*s*' with its own predicate. The plausible result is

(*T*) *s* is *T* if and only if *p*.

What we require of a theory of meaning for a language *L* is that without appeal to any (further) semantical notions it place enough restrictions on the predicate 'is *T*' to entail all sentences got from schema

T when '*s*' is replaced by a structural description of a sentence of *L* and '*p*' by that sentence.

Any two predicates satisfying this condition have the same extension,[6] so if the metalanguage is rich enough, nothing stands in the way of putting what I am calling a theory of meaning into the form of an explicit definition of a predicate 'is *T*'. But whether explicitly defined or recursively characterized, it is clear that the sentences to which the predicate 'is *T*' applies will be just the true sentences of *L*, for the condition we have placed on satisfactory theories of meaning is in essence Tarski's Convention *T* that tests the adequacy of a formal semantical definition of truth.[7]

The path to this point has been tortuous, but the conclusion may be stated simply: a theory of meaning for a language *L* shows 'how the meanings of sentences depend upon the meanings of words' if it contains a (recursive) definition of truth-in-L. And, so far at least, we have no other idea how to turn the trick. It is worth emphasizing that the concept of truth played no ostensible role in stating our original problem. That problem, upon refinement, led to the view that an adequate theory of meaning must characterize a predicate meeting certain conditions. It was in the nature of a discovery that such a predicate would apply exactly to the true sentences. I hope that what I am saying may be described in part as defending the philosophical importance of Tarski's semantical concept of truth. But my defence is only distantly related, if at all, to the question whether the concept Tarski has shown how to define is the (or a) philosophically interesting conception of truth, or the question whether Tarski has cast any light on the ordinary use of such words as 'true' and 'truth'. It is a misfortune that dust from futile and confused battles over these questions has prevented those with a theoretical interest in language—philosophers, logicians, psychologists, and linguists alike—from seeing in the semantical concept of truth (under whatever name) the sophisticated and powerful foundation of a competent theory of meaning.

There is no need to suppress, of course, the obvious connection between a definition of truth of the kind Tarski has shown how to construct, and the concept of meaning. It is this: the definition works by giving necessary and sufficient conditions for the truth of every sentence, and to give truth conditions is a way of giving the meaning of a sentence.

[6] Assuming, of course, that the extension of these predicates is limited to the sentences of *L*.

[7] A. Tarski, 'The Concept of Truth in Formalized Languages', in J. H. Woodger (trans.), *Logic, Semantics, Metamathematics* (Oxford: Oxford University Press, 1956), 152–278.

To know the semantic concept of truth for a language is to know what it is for a sentence—any sentence—to be true, and this amounts, in one good sense we can give to the phrase, to understanding the language. This at any rate is my excuse for a feature of the present discussion that is apt to shock old hands; my freewheeling use of the word 'meaning', for what I call a theory of meaning has after all turned out to make no use of meanings, whether of sentences or of words. Indeed, since a Tarski-type truth definition supplies all we have asked so far of a theory of meaning, it is clear that such a theory falls comfortably within what Quine terms the 'theory of reference' as distinguished from what he terms the 'theory of meaning'. So much to the good for what I call a theory of meaning, and so much, perhaps, against my so calling it.[8]

A theory of meaning (in my mildly perverse sense) is an empirical theory, and its ambition is to account for the workings of a natural language. Like any theory, it may be tested by comparing some of its consequences with the facts. In the present case this is easy, for the theory has been characterized as issuing in an infinite flood of sentences each giving the truth conditions of a sentence; we only need to ask, in sample cases, whether what the theory avers to be the truth conditions for a sentence really are. A typical test case might involve deciding whether the sentence 'Snow is white' *is* true if and only if snow is white. Not all cases will be so simple (for reasons to be sketched), but it is evident that this sort of test does not invite counting noses. A sharp conception of what constitutes a theory in this domain furnishes an exciting context for raising deep questions about when a theory of language is correct and how it is to be tried. But the difficulties are theoretical, not practical. In application, the trouble is to get a theory that comes close to working; anyone can tell whether it is right.[9] One can see why this is so. The theory reveals nothing new about the conditions under which an individual sentence is true; it does not make those conditions any clearer than the sentence itself does. The work of the theory is in relating the known truth conditions of each sentence to those

[8] But Quine may be quoted in support of my usage: 'in point of meaning . . . a word may be said to be determined to whatever extent the truth or falsehood of its contexts is determined' ('Truth by Convention', in *The Ways of Paradox* (New York: Random House, 1966), 82). Since a truth definition determines the truth value of every sentence in the object language (relative to a sentence in the metalanguage), it determines the meaning of every word and sentence. This would seem to justify the title Theory of Meaning.

[9] To give a single example: it is clearly a count in favour of a theory that it entails '"Snow is white" is true if and only if snow is white'. But to contrive a theory that entails this (and works for all related sentences) is not trivial. I do not know a wholly satisfactory theory that succeeds with this very case (the problem of 'mass terms').

aspects ('words') of the sentence that recur in other sentences, and can be assigned identical roles in other sentences. Empirical power in such a theory depends on success in recovering the structure of a very complicated ability—the ability to speak and understand a language. We can tell easily enough when particular pronouncements of the theory comport with our understanding of the language; this is consistent with a feeble insight into the design of the machinery of our linguistic accomplishments.

The remarks of the last paragraph apply directly only to the special case where it is assumed that the language for which truth is being characterized is part of the language used and understood by the characterizer. Under these circumstances, the framer of a theory will as a matter of course avail himself when he can of the built-in convenience of a metalanguage with a sentence guaranteed equivalent to each sentence in the object language. Still, this fact ought not to con us into thinking a theory any more correct that entails '"Snow is white" is true if and only if snow is white' than one that entails instead:

(*S*) 'Snow is white' is true if and only if grass is green,

provided, of course, we are as sure of the truth of (*S*) as we are of that of its more celebrated predecessor. Yet (*S*) may not encourage the same confidence that a theory that entails it deserves to be called a theory of meaning.

The threatened failure of nerve may be counteracted as follows. The grotesqueness of (*S*) is in itself nothing against a theory of which it is a consequence, provided the theory gives the correct results for every sentence (on the basis of its structure, there being no other way). It is not easy to see how (*S*) could be party to such an enterprise, but if it were—if, that is, (*S*) followed from a characterization of the predicate 'is true' that led to the invariable pairing of truths with truths and falsehoods with falsehoods—then there would not, I think, be anything essential to the idea of meaning that remained to be captured.[10]

What appears to the right of the biconditional in sentences of the form '*s* is true if and only if *p*' when such sentences are consequences of a theory of truth plays its role in determining the meaning of *s* not by pretending synonymy but by adding one more brush-stroke to the picture which, taken as a whole, tells what there is to know of the

[10] Critics have often failed to notice the essential proviso mentioned in this paragraph. The point is that (*S*) could not belong to any reasonably simple theory that also gave the right truth conditions for 'That is snow' and 'This is white'. (See the discussion of indexical expressions below.) [Footnote added in 1982.]

meaning of *s*; this stroke is added by virtue of the fact that the sentence that replaces '*p*' is true if and only if *s* is.

It may help to reflect that (*S*) is acceptable, if it is, because we are independently sure of the truth of 'Snow is white' and 'Grass is green'; but in cases where we are unsure of the truth of a sentence, we can have confidence in a characterization of the truth predicate only if it pairs that sentence with one we have good reason to believe equivalent. It would be ill advised for someone who had any doubts about the colour of snow or grass to accept a theory that yielded (*S*), even if his doubts were of equal degree, unless he thought the colour of the one was tied to the colour of the other.[11] Omniscience can obviously afford more bizarre theories of meaning than ignorance; but then, omniscience has less need of communication.

It must be possible, of course, for the speaker of one language to construct a theory of meaning for the speaker of another, though in this case the empirical test of the correctness of the theory will no longer be trivial. As before, the aim of theory will be an infinite correlation of sentences alike in truth. But this time the theory-builder must not be assumed to have direct insight into likely equivalences between his own tongue and the alien. What he must do is find out, however he can, what sentences the alien holds true in his own tongue (or better, to what degree he holds them true). The linguist then will attempt to construct a characterization of truth-for-the-alien which yields, so far as possible, a mapping of sentences held true (or false) by the alien on to sentences held true (or false) by the linguist. Supposing no perfect fit is found, the residue of sentences held true translated by sentences held false (and vice versa) is the margin for error (foreign or domestic). Charity in interpreting the words and thoughts of others is unavoidable in another direction as well: just as we must maximize agreement, or risk not making sense of what the alien is talking about, so we must maximize the self-consistency we attribute to him, on pain of not understanding *him*. No single principle of optimum charity emerges; the constraints therefore determine no single theory. In a theory of radical translation (as Quine calls it) there is no completely disentangling questions of what

[11] This paragraph is confused. What it should say is that sentences of the theory are empirical generalizations about speakers, and so must not only be true but also lawlike. (*S*) presumably is not a law, since it does not support appropriate counterfactuals. It's also important that the evidence for accepting the (time and speaker relativized) truth conditions for 'That is snow' is based on the causal connection between a speaker's assent to the sentence and the demonstrative presentation of snow. For further discussion, see 'Reply to Foster', in Davidson, *Inquiries into Truth and Interpretation*, 171–9. [Footnote added in 1982.]

the alien means from questions of what he believes. We do not know what someone means unless we know what he believes; we do not know what someone believes unless we know what he means. In radical interpretation we are able to break into this circle, if only incompletely, because we can sometimes tell that a person accedes to a sentence we do not understand.[12]

In the past few pages I have been asking how a theory of meaning that takes the form of a truth definition can be empirically tested, and have blithely ignored the prior question whether there is any serious chance such a theory can be given for a natural language. What are the prospects for a formal semantical theory of a natural language? Very poor, according to Tarski; and I believe most logicians, philosophers of language, and linguists agree.[13] Let me do what I can to dispel the pessimism. What I can in a general and programmatic way, of course, for here the proof of the pudding will certainly be in the proof of the right theorems.

Tarski concludes the first section of his classic essay on the concept of truth in formalized languages with the following remarks, which he italicizes:

> . . . *The very possibility of a consistent use of the expression 'true sentence' which is in harmony with the laws of logic and the spirit of everyday language seems to be very questionable, and consequently the same doubt attaches to the possibility of constructing a correct definition of this expression.* (165)

Late in the same essay, he returns to the subject:

> . . . the concept of truth (as well as other semantical concepts) when applied to colloquial language in conjunction with the normal laws of logic leads inevitably to confusions and contradictions. Whoever wishes, in spite of all difficulties, to pursue the semantics of colloquial language with the help of exact methods will be driven first to undertake the thankless task of a reform of this language. He will find it necessary to define its structure, to overcome the ambiguity of the terms which occur in it, and finally to split the language into a series of languages of

[12] This sketch of how a theory of meaning for an alien tongue can be tested obviously owes its inspiration to Quine's account of radical translation in Chapter II of *Word and Object* (Cambridge, Mass.: MIT Press, 1960). In suggesting that an acceptable theory of radical translation take the form of a recursive characterization of truth, I go beyond Quine. Toward the end of this essay, in the discussion of demonstratives, another strong point of agreement will turn up.

[13] So far as I am aware, there has been very little discussion of whether a formal truth definition can be given for a natural language. But in a more general vein, several people have urged that the concepts of formal semantics be applied to natural language. See, for example, the contributions of Yehoshua Bar-Hillel and Evert Beth in P. A. Schilpp (ed.), *The Philosophy of Rudolph Carnap* (La Salle, Ill.: Open Court, 1963), and Bar-Hillel's 'Logical Syntax and Semantics', *Language*, 30 (1954), 230–7.

greater and greater extent, each of which stands in the same relation to the next in which a formalized language stands to its metalanguage. It may, however be doubted whether the language of everyday life, after being 'rationalized' in this way, would still preserve its naturalness and whether it would not rather take on the characteristic features of the formalized languages. (267)

Two themes emerge: that the universal character of natural languages leads to contradiction (the semantic paradoxes), and that natural languages are too confused and amorphous to permit the direct application of formal methods. The first point deserves a serious answer, and I wish I had one. As it is, I will say only why I think we are justified in carrying on without having disinfected this particular source of conceptual anxiety. The semantic paradoxes arise when the range of the quantifiers in the object language is too generous in certain ways. But it is not really clear how unfair to Urdu or to Wendish it would be to view the range of their quantifiers as insufficient to yield an explicit definition of 'true-in-Urdu' or 'true-in-Wendish'. Or, to put the matter in another, if not more serious way, there may in the nature of the case always be something we grasp in understanding the language of another (the concept of truth) that we cannot communicate to him. In any case, most of the problems of general philosophical interest arise within a fragment of the relevant natural language that may be conceived as containing very little set theory. Of course these comments do not meet the claim that natural languages are universal. But it seems to me that this claim, now that we know such universality leads to paradox, is suspect.

Tarski's second point is that we would have to reform a natural language out of all recognition before we could apply formal semantical methods. If this is true, it is fatal to my project, for the task of a theory of meaning as I conceive it is not to change, improve, or reform a language, but to describe and understand it. Let us look at the positive side. Tarski has shown the way to giving a theory for interpreted formal languages of various kinds; pick one as much like English as possible. Since this new language has been explained in English and contains much English we not only may, but I think must, view it as part of English for those who understand it. For this fragment of English we have, *ex hypothesi*, a theory of the required sort. Not only that, but in interpreting this adjunct of English in old English we necessarily gave hints connecting old and new. Wherever there are sentences of old English with the same truth conditions as sentences in the adjunct we may extend the theory to cover them. Much of what is called for is to mechanize as far as possible what we now do by art when we put ordinary English into one or another canonical notation. The point is not that canonical notation is better than the rough original idiom, but rather that if we know what idiom the canonical notation is canonical *for*, we have as good a theory for the idiom as for its kept companion.

Philosophers have long been at the hard work of applying theory to ordinary language by the device of matching sentences in the vernacular with sentences for which they have a theory. Frege's massive contribution was to show how 'all', 'some', 'every', 'each', 'none', and associated pronouns, in some of their uses, could be tamed; for the first time, it was possible to dream of a formal semantics for a significant part of a natural language. This dream came true in a sharp way with the work of Tarski. It would be a shame to miss the fact that as a result of these two magnificent achievements, Frege's and Tarski's, we have gained a deep insight into the structure of our mother tongues. Philosophers of a logical bent have tended to start where the theory was and work out towards the complications of natural language. Contemporary linguists, with an aim that cannot easily be seen to be different, start with the ordinary and work toward a general theory. If either party is successful, there must be a meeting. Recent work by Chomsky and others is doing much to bring the complexities of natural languages within the scope of serious theory. To give an example: suppose success in giving the truth conditions for some significant range of sentences in the active voice. Then with a formal procedure for transforming each such sentence into a corresponding sentence in the passive voice, the theory of truth could be extended in an obvious way to this new set of sentences.[14]

One problem touched on in passing by Tarski does not, at least in all its manifestations, have to be solved to get ahead with theory: the existence in natural languages of 'ambiguous terms'. As long as ambiguity does not affect grammatical form, and can be translated, ambiguity for ambiguity, into the metalanguage, a truth definition will

[14] The *rapprochement* I prospectively imagine between transformational grammar and a sound theory of meaning has been much advanced by a recent change in the conception of transformational grammar described by Chomsky in the article referred to above (note 5). The structures generated by the phrase-structure part of the grammar, it has been realized for some time, are those suited to semantic interpretation; but this view is inconsistent with the idea, held by Chomsky until recently, that recursive operations are introduced only by the transformation rules. Chomsky now believes the phrase-structure rules are recursive. Since languages to which formal semantic methods directly and naturally apply are ones for which a (recursive) phrase-structure grammar is appropriate, it is clear that Chomsky's present picture of the relation between the structures generated by the phrase-structure part of the grammar, and the sentences of the language, is very much like the picture many logicians and philosophers have had of the relation between the richer formalized languages and ordinary language. (In these remarks I am indebted to Bruce Vermazen.)

not tell us any lies. The chief trouble, for systematic semantics, with the phrase 'believes that' in English lies not in its vagueness, ambiguity, or unsuitability for incorporation in a serious science: let our metalanguage be English, and all *these* problems will be carried without loss or gain into the metalanguage. But the central problem of the logical grammar of 'believes that' will remain to haunt us.

The example is suited to illustrating another, and related, point, for the discussion of belief sentences has been plagued by failure to observe a fundamental distinction between tasks: uncovering the logical grammar or form of sentences (which is in the province of a theory of meaning as I construe it), and the analysis of individual words or expressions (which are treated as primitive by the theory). Thus Carnap, in the first edition of *Meaning and Necessity*, suggested we render 'John believes that the earth is round' as 'John responds affirmatively to "the earth is round" as an English sentence'. He gave this up when Mates pointed out that John might respond affirmatively to one sentence and not to another no matter how close in meaning.[15] But there is a confusion here from the start. The semantic structure of a belief sentence, according to this idea of Carnap's, is given by a three-place predicate with places reserved for expressions referring to a person, a sentence, and a language. It is a different sort of problem entirely to attempt an analysis of this predicate, perhaps along behaviouristic lines. Not least among the merits of Tarski's conception of a theory of truth is that the purity of method it demands of us follows from the formulation of the problem itself, not from the self-imposed restraint of some adventitious philosophical puritanism.

I think it is hard to exaggerate the advantages to philosophy of language of bearing in mind this distinction between questions of logical form or grammar, and the analysis of individual concepts. Another example may help advertise the point.

If we suppose questions of logical grammar settled, sentences like 'Bardot is good' raise no special problems for a truth definition. The deep differences between descriptive and evaluative (emotive, expressive, etc.) terms do not show here. Even if we hold there is some important sense in which moral or evaluative sentences do not have a truth value (for example, because they cannot be verified), we ought not to boggle at ' "Bardot is good" is true if and only if Bardot is good'; in a theory of truth, this consequence should follow with the rest, keeping track, as must be done, of the semantic location of such sentences in the

[15] B. Mates, 'Synonymity', in L. Linsky (ed.), *Semantics and the Philosophy of Language* (Urbana, Ill.: University of Illinois Press, 1952), 111–36.

language as a whole—of their relation to generalizations, their role in such compound sentences as 'Bardot is good and Bardot is foolish', and so on. What is special to evaluative words is simply not touched: the mystery is transferred from the word 'good' in the object language to its translation in the metalanguage.

But 'good' as it features in 'Bardot is a good actress' is another matter. The problem is not that the translation of this sentence is not in the metalanguage—let us suppose it is. The problem is to frame a truth definition such that '"Bardot is a good actress" is true if and only if Bardot is a good actress'—and all other sentences like it—are consequences. Obviously 'good actress' does not mean 'good and an actress'. We might think of taking 'is a good actress' as an unanalysed predicate. This would obliterate all connection between 'is a good actress' and 'is a good mother', and it would give us no excuse to think of 'good', in these uses, as a word or semantic element. But worse, it would bar us from framing a truth definition at all, for there is no end to the predicates we would have to treat as logically simple (and hence accommodate in separate clauses in the definition of satisfaction): 'is a good companion to dogs', 'is a good 28-years old conversationalist', and so forth. The problem is not peculiar to the case: it is the problem of attributive adjectives generally.

It is consistent with the attitude taken here to deem it usually a strategic error to undertake philosophical analysis of words or expressions which is not preceded by or at any rate accompanied by the attempt to get the logical grammar straight. For how can we have any confidence in our analyses of words like 'right', 'ought', 'can', and 'obliged', or the phrases we use to talk of actions, events, and causes, when we do not know what (logical, semantical) parts of speech we have to deal with? I would say much the same about studies of the 'logic' of these and other words, and the sentences containing them. Whether the effort and ingenuity that have gone into the study of deontic logics, modal logics, imperative and erotetic logics have been largely futile or not cannot be known until we have acceptable semantic analyses of the sentences such systems purport to treat. Philosophers and logicians sometimes talk or work as if they were free to choose between, say, the truth-functional conditional and others, or free to introduce non-truth-functional sentential operators like 'Let it be the case that' or 'It ought to be the case that'. But in fact the decision is crucial. When we depart from idioms we can accommodate in a truth definition, we lapse into (or create) language for which we have no coherent semantical account—

that is, no account at all of how such talk can be integrated into the language as a whole.

To return to our main theme: we have recognized that a theory of the kind proposed leaves the whole matter of what individual words mean exactly where it was. Even when the metalanguage is different from the object language, the theory exerts no pressure for improvement, clarification, or analysis of individual words, except when, by accident of vocabulary, straightforward translation fails. Just as synonomy, as between expressions, goes generally untreated, so also synonomy of sentences, and analyticity. Even such sentences as 'A vixen is a female fox' bear no special tag unless it is our pleasure to provide it. A truth definition does not distinguish between analytic sentences and others, except for sentences that owe their truth to the presence alone of the constants that give the theory its grip on structure: the theory entails not only that these sentences are true but that they will remain true under all significant rewritings of their non-logical parts. A notion of logical truth thus given limited application, related notions of logical equivalence and entailment will tag along. It is hard to imagine how a theory of meaning could fail to read a logic into its object language to this degree; and to the extent that it does, our intuitions of logical truth, equivalence, and entailment may be called upon in constructing and testing the theory.

I turn now to one more, and very large, fly in the ointment: the fact that the same sentence may at one time or in one mouth be true and at another time or in another mouth be false. Both logicians and those critical of formal methods here seem largely (though by no means universally) agreed that formal semantics and logic are incompetent to deal with the disturbances caused by demonstratives. Logicians have often reacted by downgrading natural language and trying to show how to get along without demonstratives; their critics react by downgrading logic and formal semantics. None of this can make me happy: clearly demonstratives cannot be eliminated from a natural language without loss or radical change, so there is no choice but to accommodate theory to them.

No logical errors result if we simply treat demonstratives as constants;[16] neither do any problems arise for giving a semantic truth definition. '"I am wise" is true if and only if I am wise', with its bland ignoring of the demonstrative element in 'I' comes off the assembly line along with '"Socrates is wise" is true if and only if Socrates is wise' with

[16] See W. V. Quine, *Methods of Logic* (New York: Holt, 1950), 8.

its bland indifference to the demonstrative element in 'is wise' (the tense).

What suffers in this treatment of demonstratives is not the definition of a truth predicate, but the plausibility of the claim that what has been defined is truth. For this claim is acceptable only if the speaker and circumstances of utterance of each sentence mentioned in the definition is matched by the speaker and circumstances of utterance of the truth definition itself. It could also be fairly pointed out that part of understanding demonstratives is knowing the rules by which they adjust their reference to circumstance; assimilating demonstratives to constant terms obliterates this feature. These complaints can be met, I think, though only by a fairly far-reaching revision in the theory of truth. I shall barely suggest how this could be done, but bare suggestion is all that is needed: the idea is technically trivial, and in line with work being done on the logic of the tenses.[17]

We could take truth to be a property, not of sentences, but of utterances, or speech acts, or ordered triples of sentences, times, and persons; but it is simplest just to view truth as a relation between a sentence, a person, and a time. Under such treatment, ordinary logic as now read applies as usual, but only to sets of sentences relativized to the same speaker and time; further logical relations between sentences spoken at different times and by different speakers may be articulated by new axioms. Such is not my concern. The theory of meaning undergoes a systematic but not puzzling change; corresponding to each expression with a demonstrative element there must in the theory be a phrase that relates the truth conditions of sentences in which the expression occurs to changing times and speakers. Thus the theory will entail sentences like the following:

> 'I am tired' is true as (potentially) spoken by p at t if and only if p is tired at t.
>
> 'That book was stolen' is true as (potentially) spoken by p at t if and only if the book demonstrated by p at t is stolen prior to t.[18]

Plainly, this course does not show how to eliminate demonstratives; for example, there is no suggestion that 'the book demonstrated by the speaker' can be substituted ubiquitously for 'that book' *salva veritate*.

[17] This claim has turned out to be naïvely optimistic. For some serious work on the subject, see S. Weinstein, 'Truth and Demonstratives', *Noûs*, 8 (1974), 179–84. [Note added in 1982.]

[18] There is more than an intimation of this approach to demonstratives and truth in J. L. Austin, 'Truth', *Proceedings of the Aristotelian Society*, Supp. vol. 24 (1950).

The fact that demonstratives are amenable to formal treatment ought greatly to improve hopes for a serious semantics of natural language, for it is likely that many outstanding puzzles, such as the analysis of quotations or sentences about propositional attitudes, can be solved if we recognize a concealed demonstrative construction.

Now that we have relativized truth to times and speakers, it is appropriate to glance back at the problem of empirically testing a theory of meaning for an alien tongue. The essence of the method was, it will be remembered, to correlate held-true sentences with held-true sentences by way of a truth definition, and within the bounds of intelligible error. Now the picture must be elaborated to allow for the fact that sentences are true, and held true, only relative to a speaker and a time. Sentences with demonstratives obviously yield a very sensitive test of the correctness of a theory of meaning, and constitute the most direct link between language and the recurrent macroscopic objects of human interest and attention.[19]

In this essay I have assumed that the speakers of a language can effectively determine the meaning or meanings of an arbitrary expression (if it has a meaning). and that it is the central task of a theory of meaning to show how this is possible. I have argued that a characterization of a truth predicate describes the required kind of structure, and provides a clear and testable criterion of an adequate semantics for a natural language. No doubt there are other reasonable demands that may be put on a theory of meaning. But a theory that does no more than define truth for a language comes far closer to constituting a complete theory of meaning than superficial analysis might suggest; so, at least, I have urged.

Since I think there is no alternative, I have taken an optimistic and programmatic view of the possibilities for a formal characterization of a truth predicate for a natural language. But it must be allowed that a staggering list of difficulties and conundrums remains. To name a few: we do not know the logical form of counterfactual or subjunctive sentences; nor of sentences about probabilities and about causal relations; we have no good idea what the logical role of adverbs is, nor the role of attributive adjectives; we have no theory for mass terms like 'fire', 'water', and 'snow', nor for sentences about belief, perception, and intention, nor for verbs of action that imply purpose. And finally, there are all the sentences that seem not to have truth values at all: the

[19] These remarks derive from Quine's idea that 'occasion sentences' (those with a demonstrative element) must play a central role in constructing a translation manual.

imperatives, optatives, interrogatives, and a host more. A comprehensive theory of meaning for a natural language must cope successfully with each of these problems.[20]

[20] For attempted solutions to some of these problems, see Essays 6–10 of *Essays on Actions and Events* (Oxford: Oxford University Press, 1980), and Essays 6–8 of *Inquiries into Truth and Interpretation*. There is further discussion in Essays 3, 4, 9, and 10, and reference to some progress in sect. 1 of Essay 9 in the latter book.

VII

ON THE SENSE AND REFERENCE OF A PROPER NAME

JOHN McDOWELL

I

An interesting way to raise questions about the relation between language and reality is to ask: how could we state a theory knowledge of which would suffice for understanding a language? Donald Davidson has urged that a central component in such a theory would be a theory of truth, in something like the style of Tarski, for the language in question.[1] A Tarskian truth-theory entails, for each indicative sentence of the language it deals with, a theorem specifying a necessary and sufficient condition for the sentence to be true. The theorems are derivable from axioms which assign semantic properties to sentence-constituents and determine the semantic upshot of modes of combination. Now Frege held that the senses of sentences can be determined by giving truth-conditions, and that the sense of a sentence-constituent is its contribution to the senses of sentences in which it may occur.[2] The parallel is striking. It suggests a construal of Davidson's proposal as a proposal about the nature of a theory of (Fregean) sense for a language.

Tarskian truth-theories are extensional, and only minimally richer in ontology than their object languages (no need to mention possible worlds and the like). This attractive economy of resources has a consequence which is crucial to the proper understanding of the suggestion: namely that not just any theory of truth, however true, could serve as a theory of sense.

The job of a theory of sense should be to fix the content of speech-acts

From *Mind*, 86 (1977), 159–85. Reprinted by permission of Oxford University Press.
I have been reading ancestors of this essay to seminars and discussion groups since 1971: it would be impossible for me to identify all the influences on it. I should like to acknowledge special debts to Gareth Evans and David Wiggins.

[1] See Donald Davidson, 'Radical Interpretation', *Dialectica*, 27 (1973), 313.

[2] G. Frege, *The Basic Laws of Arithmetic: Exposition of the System*, trans. and ed. Montgomery Furth (Berkeley and Los Angeles: University of California Press, 1967), 89–90.

which a total theory of the language concerned would warrant ascribing to speakers.[3] Abstracting harmlessly from complications induced by indexicality and non-indicative utterances, we can put the point like this: in the case of any sentence whose utterance command of the language would make fully comprehensible as a saying—any indicative sentence—a theory of sense must fix the content of the saying which an intentional utterance of the sentence could be understood to be.

The adequacy of the total theory would turn on its acceptably imposing descriptions, reporting behaviour as performance of speech-acts of specified kinds with specified contents, on a range of potential actions—those which would constitute speech in the language—describable, antecedently, only as so much patterned emission of noise. For that systematic imposing of descriptions to be acceptable, it would have to be the case that speakers' performances of the actions thus ascribed to them were, for the most part, intelligible under those descriptions, in the light of propositional attitudes: their possession of which, in turn, would have to be intelligible, in the light of their behaviour—including, of course, their linguistic behaviour—and their environment. The point of the notion of sense—that which the content-specifying component of a total theory of that sort would be a theory of—is thus tied to our interest in the understanding of behaviour, and ultimately our interest in the understanding—the fathoming—of people. We have not properly made sense of forms of words in a language if we have not, thereby, got some way towards making sense of its speakers. If there is a pun here, it is an illuminating one.

Now to specify the content of a saying we need a sentence. A theory might have the power to give us a sentence to meet the need, for any indicative sentence of its object language, by virtue of the fact that it entailed theorems in which the needed content-specifying sentences were used to state necessary and sufficient conditions for the application, to the relevant object-language sentences, of some predicate. The fact that the used sentences specified the content of sayings potentially effected by uttering the mentioned sentences would guarantee that the predicate could, if we liked, be written 'true'; it would guarantee that the theory, with its theorems written that way, was a true theory of truth.[4] But it

[3] See 'On Sense and Reference', in Peter Geach and Max Black, *Translations from the Philosophical Writings of Gottlob Frege* (Oxford: Blackwell, 1952) (now Essay I of this volume). The sense of a sentence determines, or is, the thought expressed (p. 28, above); it would be specified in a 'that' clause (see the remark about reported speech, p. 25, above).

[4] See Donald Davidson, 'Truth and Meaning', *Synthese*, 17 (1967), 310–11 (now Essay VI of this volume).

would be the guaranteeing fact, and not the guaranteed fact, which suited the theory to serve as a theory of sense. For, given a theory guaranteed to be a true theory of truth, by its serviceability in yielding content-specifications, we could exploit the extensionality of truth-theories to derive a new, equally true theory of truth which, in spite of its truth, would not be serviceable in yielding content-specifications, and so would not serve as a theory of sense.[5]

What emerges is that serving as a theory of sense is not the same as being one, on a certain strict view of what it is to be one. It was clear anyway that a truth-theory of the sort Davidson envisages does not, in saying what it does, *state* the senses of expressions. Why should we hanker after a theory which does that mysterious thing, if a theory which does some utterly unmysterious thing instead can be made to serve the purpose? Pending a good answer to that question, there is only the mildest perversity in shifting our conception of what a theory of sense must be, so that Davidsonian theories are at least not ruled out. As positive justification we still have the striking parallel noted at the outset between the Davidsonian *Ersatz* and Frege's own ideas about the genuine article.

II

I shall restrict attention throughout to occurrences of names in straightforwardly extensional contexts in which they function as singular terms; and I shall, in this paper, ignore names with more than one bearer.

For simplicity, I shall begin by considering theories which aim to deal with (fragments of) English in English (the restriction will be removed in Section VII). In a theory of truth of that kind, names might be handled by axioms of which the following is typical:[6]

'Hesperus' stands for (denotes) Hesperus.

The role played by some such clause, in the derivation of assignments of truth-conditions to sentences in which the name occurs, would display the contribution made by the name to those truth-conditions. Given the

[5] See the Introduction to Gareth Evans and John McDowell (eds.), *Truth and Meaning* (Oxford: Clarendon Press, 1976), pp. xv–xvii.

[6] Smoothness in treatment of predicates (whose argument-places can be filled with variables as well as names) dictates, rather, the statement that each of a certain set of functions from singular terms (definite or indefinite) to objects assigns Hesperus to 'Hesperus'; but it comes to the same thing.

Fregean doctrine about sense mentioned at the beginning of this paper, this suggests that such a clause, considered as having what it says fixed by its location in a theory which yields acceptable content-specifications, gives—or, more strictly, in that context as good as gives—the sense of the name.

III

If there is to be any affinity between my use of 'sense' and Frege's of '*Sinn*' I must keep room for a distinction between sense and reference. There must be contexts where 'sense' is required and 'reference' would not do. Now clauses of the sort just exemplified specify, surely, *references (Bedeutungen)* of names, and it might be thought that the distinction has disappeared.

That thought would be wrong. Frege's notion of sense belongs with the notion of understanding, and we can get at what is involved in understanding a language by careful employment of the notion of knowledge. Recall my opening question about the nature of a theory knowledge of which would suffice for understanding a language. We can think of a theory of sense as a component of a total theory of that kind: a component which, in the context of principles adequate to determine the force (assertion, instruction, or whatever) with which particular utterances are issued, serves (as suggested in Section I) to determine the content of speech-acts performed in issuing those utterances. Semantically simple expressions would be mentioned in axioms of such a theory, designed so that knowledge of the truths they express—in the context of knowledge of enough of the rest of the theory—would suffice for understanding utterances containing those expressions. The hypothetical knowledge involved here, then, is knowledge of truths (French '*savoir*', German '*wissen*'). The reference (*Bedeutung*) of a name, on the other hand, is, in Frege's usage, its bearer—an object.[7] To know the reference of a name would be, failing an unpardonable equivocation, to know that object: acquaintance, perhaps, but in any case not knowledge of truths but, what is grammatically distinct, knowledge of things (French '*connaître*', German '*kennen*'). It is not, then, the sort of knowledge

[7] Notwithstanding Michael Dummett's assertion that Frege had two other uses of '*Bedeutung*' as well (Frege, *Philosophy of Language* (London: Duckworth, 1973), 93–4). Casual occurrences of that ordinary German word should carry no weight against the intention which is plain in Frege's official exposition of the doctrine.

which it would make sense to *state* in clauses of a theory. The grammatical distinction between knowledge of things and knowledge of truths guarantees a difference of role for 'sense' and 'reference'. Without putting that difference at risk, we can claim that a clause which does no more than state—in a suitable way—what the reference of an expression is may nevertheless give—or as good as give—that expression's sense.

This grammatical way of distinguishing sense and reference promises to free us from any need to worry about the ontological status of senses. As far as names are concerned, the ontology of a theory of sense, on the present suggestion, need not exceed the names and their bearers. To construe knowledge of the sense of an expression (knowledge of a truth) as, at some different level, knowledge of (perhaps acquaintance with) an entity (the sense of the expression) seems from this perspective, gratuitous.[8]

Verbal nouns like 'reference' (and '*Bedeutung*') have a curious grammatical property which facilitates temptation to the equivocation alluded to above. We would do well to immunize ourselves to the temptation by exposing its source. A phrase of the form 'the reference of *x*' can be understood as equivalent to the corresponding phrase of the form 'what *x* refers to', either (i) in the sense in which 'what' amounts to 'that which' (which yields the official Fregean use of '*Bedeutung*') or (ii) in the sense in which 'what' is an interrogative pronoun. In this second sense, 'what *x* refers to' gives the form of an indirect question, something suitable to follow 'know' where knowledge of truths is what is meant. Knowledge of the *reference* of a name, in this second (non-Fregean) sense, could reasonably be held to be knowledge which, in the context of appropriate further knowledge not itself involving the name, would suffice for understanding utterances containing the name—that is, precisely, knowledge of its *sense*. It is presumably just this slipperiness of the word 'reference' which motivates the introduction of the term of art 'referent'.

The possibility of equating a distinction between sense and (Fregean) reference with a distinction between (non-Fregean) reference and referent may, for those who are at home in the latter idiom, make it easier to see how crediting names with senses (to persist, with the plural, in an ontologically incautious formulation) is not necessarily crediting them with anything like connotation or descriptive meaning.

[8] I am not suggesting that Frege attained this perspective himself.

IV

Frege's distinction was first put to use to solve a problem about identity sentences. Altered so as to involve undisputable instances of ordinary proper names, the question is this: how could someone possess knowledge sufficient to understand a sentence like 'Hesperus is Phosphorus' without thereby knowing, already, that the sentence is true?[9] That question can be answered in the context of a theory which treats names in the austere way suggested above. The two names would be treated, in a theory of the sort I am envisaging, by clauses to this effect:

'Hesperus' stands for Hesperus.
'Phosphorus' stands for Phosphorus.

Since Hesperus *is* Phosphorus, the right-hand sides of these clauses can of course be interchanged *salva veritate*. But it does not follow, from the truth of the results, that they would be equally serviceable in a theory of sense (cf. Section I). The idea was that knowledge that 'Hesperus' stands for Hesperus would suffice, in the context of suitable other knowledge not directly involving the name, for understanding utterances containing 'Hesperus'; and similarly with knowledge that 'Phosphorus' stands for Phosphorus and utterances containing 'Phosphorus'. Now if someone knows that 'Hesperus' stands for Hesperus and that 'Phosporus' stands for Phosphorus, it does not follow that he knows either that 'Hesperus' stands for Phosphorus or that 'Phosphorus' stands for Hesperus. And for it to seem that knowledge sufficient for understanding the sentence would, of itself, suffice for knowing its truth, one or the other of those implications would have to be thought to hold.[10]

The point does not turn essentially on treating the *deductive apparatus* of the envisaged theory of sense (in particular, its axioms) as spelling out hypothetical knowledge (knowledge which would suffice for understanding). The inadequacy of a theory which used just one of the two names on the right-hand sides of its clauses for both would show in its *consequences*, the theorems assigning truth-conditions to sentences. We would have to use those consequences in fixing the content of sayings, to be found intelligible in terms of propositional attitudes, our specification of whose content would need to be fixed partly by our interpretation of the sayings. And some utterances of 'Hesperus is Phosphorus' would be

[9] See above, pp. 23–4.

[10] Michael Dummett alludes to an earlier version of this paper, in 'What is a Theory of Meaning?' (in Samuel Guttenplan (ed.), *Mind and Language* (Oxford: Clarendon Press, 1975)), 122. This paragraph should make it clear that he has indeed misinterpreted me, as he conjectures (in his Appendix) at 126.

unintelligible—we would not be able to see their point—if we were reduced to regarding them as expressing the belief that, say, Hesperus is Hesperus.[11] What we have here is a glimpse of the way in which, by requiring the theory's consequences to help us make sense of speakers of the language, we force ourselves to select among the multiplicity of true theories of truth. Bearing directly on the theorems, the requirement bears indirectly on the deductive apparatus which generates them. The failures of substitution exploited above, within contexts which specify hypothetical knowledge, simply make vivid how the indirect requirement operates.

V

According to Michael Dummett's reconstruction, Frege's own view was that the sense of a name is a criterion for, or way of, recognizing or identifying an object as its bearer.[12] I have so far left ways of recognizing objects entirely out of account. It is noteworthy how much of Frege it has nevertheless been possible to preserve. Frege's own examples undeniably manifest a richer conception of how sense might be represented; but the suspicion arises that that might be an unnecessary excrescence, rather than—what Dummett evidently takes it to be—something essential to any elaboration of Frege's basic ideas about sense.

What is the point of the notion of sense? According to Dummett's attractive suggestion, it is to capture (in part) a notion of meaning which makes it true that a theory of meaning is a theory of understanding.[13] A theory of understanding is just what I have been thinking of a theory of a language as being. In my terms, then, the issue is this. Within the present restricting assumptions, would knowledge that 'Hesperus' stands for Hesperus—in the context of suitable knowledge about other expressions, and suitable knowledge about the forces with which utterances may be issued—suffice for understanding utterances containing 'Hesperus'? Or would one require, rather, knowledge to the effect that

[11] What about a theory which used both names, but interchanged them? The inadequacy of this will emerge in Section IX.

[12] Dummett, *Frege*, 95 ff. Whether the way of identifying is expressible in words or not seems unimportant, *pace* Dummett, who makes much of the fact that it need not be so expressible, as against those who attack Frege as a description theorist. The profound unsatisfactoriness of treating the sense of a name as essentially descriptive (see Section VIII below) is not cured by allowing that there need be no explicitly descriptive *expression* which expresses that sense.

[13] Dummett, *Frege*, 92–3.

the bearer of the name may be recognized or identified thus and so? Now patently this second, stronger, requirement, interpreted in any ordinary way, insists on more than would suffice; for it insists on more than, in some cases, does suffice. One can have the ability to tell that a seen object is the bearer of a familiar name without having the slightest idea *how* one recognizes it. The presumed mechanism of recognition might be neural machinery—its operations quite unknown to its possessor.[14]

Understanding a language involves knowing, on occasion, what speakers of it are doing, under descriptions which report their behaviour as speech-acts of specified kinds with specified contents. It helps to picture a possessor and a non-possessor of the state involved being subjected together to speech in the language. Assuming he is awake and attentive, the one will know truths expressible by the application of such descriptions; the other will not. Certain information is made available to both, in their shared sensory experience; certain further information is possessed only by one. Now a theory of a language was described above (Section I) as warranting systematic imposition of interpreting descriptions on the range of potential behaviour which would constitute speech in the language, thought of as describable, in advance of receiving the interpreting descriptions, only as emission of noise. Such a theory, then, would have the following deductive power: given a suitable formulation of the information made available to both the possessor and the non-possessor of the state of understanding on any of the relevant potential occasions, it would permit derivation of the information which the possessor of the state would be distinguished by having. The ability to comprehend heard speech is an information-processing capacity, and the theory would describe it by articulating in detail the relation, which defines the capacity, between input information and output information.

In order to acquire an information-processing capacity with the right input–output relation, it would suffice to get to know the theory; then one could move from input information to output information, on any of the relevant occasions, by explicit deduction. It does not follow that to have such a capacity is to know any such theory. Nor, in any ordinary sense of 'know', is it usually true. Comprehension of speech in a familiar language is a matter of unreflective perception, not the bringing to bear of a theory.

[14] Neural, not psychological machinery: cf. Dummett's revealing misconception, 'What is a Theory of Meaning?', p. 122. I mean to be denying what Dummett (*Frege*, 102–3) thinks uncontroversial, namely that a speaker always has 'a route that he uses' for getting from a name to its bearer.

It is important not to be misled into a bad defence of the richer conception of the sense of a name. Certainly, it may be said, understanding a language does not consist in *explicit* knowledge of a theory. But we are not precluded, by that concession, from saying—as Dummett indeed does[15]—that understanding a language consists in *implicit* knowledge of a theory. And employment of the concept of implicit or tacit knowledge in the recent history of linguistics[16] might make one suppose that the point of this view of understanding lies in dissatisfaction with the unambitious aim of merely *describing* the state. The attraction of the notion of implicit knowledge, one might suppose, lies in its promise to permit us, more enterprisingly, to *explain* exercises of the capacity involved, in terms of a postulated inner mechanism. The workings of the mechanism are to be thought of as implicit counterparts of the explicit operations, with an explicitly known theory, whereby we might simulate the behaviour we view as the mechanism's external manifestations. Now a mechanism whose workings include an implicit counterpart to explicitly employing a method for recognizing or identifying the bearer of a name would seem to afford more explanation than an alternative mechanism constituted by implicit knowledge of an austere theory of sense. There is, with a rich theory, a specific operation of the mechanism to account for a person's getting on to the right object, and the austere conception yields nothing of the sort. And, on this view, it is beside the point to appeal to those facts about what speakers explicitly know which were exploited, above, in the argument that the rich conception insists on more knowledge than would suffice.

No such defence of the rich conception is available to Frege, as Dummett interprets him. Expounding Frege, Dummett connects the notion of sense, as remarked above, with the notion of understanding. The notion of understanding is a psychological notion, and there is the threat of an objection to using it in this context, on the ground that it infects the notion of sense with the psychologism which Frege detested. Dummett meets the threat, on Frege's behalf, precisely by denying that the notion of sense is to be thought of as employed in the construction of a purportedly explanatory mechanism (*Frege*, p. 681):

A model of sense is not a description of some hypothesized psychological mechanism . . . A model for the sense of a word of some particular kind does not

[15] For a clear statement, see Dummett, 'What is a Theory of Meaning? (II)', in Evans and McDowell, *Truth and Meaning*, 70.

[16] See e.g. Christina Graves and others, 'Tacit Knowledge', *Journal of Philosophy*, 70 (1973), 318.

seek to explain *how* we are able to use the word as we do: it simply forms part of an extended description of what that use consists in.

The point is not just *ad hominem*, since a Fregean detestation of such psychologism is well placed. There is no merit in a conception of the mind which permits us to speculate about its states, conceived as states of a hypothesized mechanism, with a breezy lack of concern for facts about explicit awareness. Postulation of implicit knowledge for such allegedly explanatory purposes sheds not scientific light but philosophical darkness.[17]

It is certainly true that psychological explanations of behaviour are central in the conception of a theory of a language outlined above (Section I). But their purpose is to confirm the descriptive adequacy of a theory, not to put an explanatory mechanism through its paces. The demand is that we should be able to see, in enough cases, why speakers might think fit to act in the ways in which, by application of the theory, they are described as intentionally acting—the point being to underpin our confidence that they are indeed acting intentionally in those ways. That is quite distinct from demanding explanations of how speakers arrive at knowledge of what others are doing, under those descriptions; or how they contrive to embody actions which are intentional under those descriptions in their own verbal behaviour.

Hostility to psychologism, then, is not hostility to the psychological. It is of the utmost importance to distinguish rejection of psychologism, as characterized in the above quotation from Dummett, from anti-mentalism of the sort which issues from Skinnerian behaviourism. Indeed one of the chief objections to the psychologistic postulation of implicit knowledge stems from a concern that the notion of the inner life, the life of the mind, not be made unrecognizable.[18] Hypothesized mechanisms are not the way to save from behaviourist attack the indispensable thought that all is not dark within. We get no authentic and satisfying conception of the mind from either of these philistine extremes.

If Frege meant the notion of sense to figure in theories without the explanatory aspirations of psychologism, why did he hold the richer conception of the sense of a name? Partly, perhaps, because of the metaphorical form in which his treatment of the 'Hesperus'–'Phosphorus'

[17] The points of Thomas Nagel, 'The Boundaries of Inner Space', *Journal of Philosophy*, 66 (1969), 452, and Stephen P. Stich, 'What Every Speaker Knows', *Philosophical Review*, 80 (1971), 476, have not been adequately answered.

[18] Self-styled Cartesians in modern linguistics are in this respect notably unfaithful to Descartes himself.

puzzle presented itself to him. The two names must not differ merely as objects—they must not be merely phonetic or orthographic variants for each other—if a genuine puzzle is to arise; they must differ also in their manner of presentation of the object which both present, and therein lies the solution.[19] Now, this metaphor of manners of presentation can be interpreted in the context of the austere conception. Difference in sense between 'Hesperus' and 'Phosphorus' lies in the fact that the clauses in the theory of sense which specify the object presented by the names are constrained to present it in the ways in which the respective names present it. They meet this constraint—surely infallibly—by actually using the respective names. But it takes subtlety to find the metaphor thus already applicable; it can easily seem to necessitate something more like a description theory of names. (Any tendency to find a richer interpretation of the metaphor natural would be reinforced by the view of thought to be discussed in Section VIII below.[20])

VI

Someone might be tempted to argue against the austere treatment of names in a theory of sense on the following lines. Knowledge that 'Hesperus' stands for Hesperus is too easily acquired for it to suffice, even in the context of appropriate further knowledge, for understanding utterances which contain the name. To know that 'Hesperus' stands for Hesperus, one needs no more than the merely syntactic knowledge that the expression is a name, together with mastery of a general trick of dropping quotation marks.

The argument is worthless. Even adding the knowledge, hardly syntactic, that the name is not bearerless, what those easily acquired accomplishments suffice for is not knowledge that 'Hesperus' stands for Hesperus, but knowledge that the sentence '"Hesperus" stands for Hesperus' expresses a truth. What we were interested in was the former state, knowledge of the truth which the sentence expresses, and not the latter.[21]

What exactly does the distinction amount to? Recall that we are concerned with these states of knowledge, not as actually possessed by all competent speakers, but as such that someone who possessed them

[19] See above, p. 24.

[20] There is also the wish (not satisfied in the austere conception) that differences in sense should not reflect but explain failures of substitution: see the last paragaph of Section IX.

[21] See Dummett, 'What is a Theory of Meaning?', 106–7.

would be able to use them in order to arrive by inference at the knowledge about particular speech-acts which a fluent hearer acquires by unreflective perception. A clause like '"Hesperus" stands for Hesperus', on my account, would figure in a theory which, for speech-acts in which the name was uttered, warranted specifying their content by means of sentences in which the name was used, that is, sentences which mentioned the planet. Such a clause would do no work in the description of a linguistic capacity actually possessed by a given speaker—knowledge of what it says would play no part in duplicating, by explicit employment of a theory, anything which he could do without reflection—unless he showed an ability to use the name, or respond intelligently (with understanding) to uses of the name on the part of others, in speech-acts construable as being about the planet. Someone whose knowledge about 'Hesperus' was limited to its being a name with a bearer would simply not be enabled thereby to behave in those ways.

If we are to find such an ability in a person, we must be able to find the relevant speech behaviour, and responses to speech behaviour, intelligible in terms of propositional attitudes possessed by him, in the specification of whose content, again, the planet would presumably need to be mentioned. There is considerable plausibility in the idea that, if we are to be able to find in a person any propositional attitudes at all about an object, we must be able to find in him some beliefs about it. If that is right, these considerations capture what looks like a grain of truth in description theories of the sense of names: a person who knows the sense of a name must have some beliefs about its bearer. But that does not amount to a justification for a less austere treatment of the name in a theory of sense. The concession envisaged is that the person must have some *beliefs*—possibly sketchy, possibly false—about the object; not that he must *know* truths about it, sufficiently full to be true of it alone, and thus capable of generating a definite description which could replace the used name in the relevant clause of the theory of sense.

VII

At this point I shall lift the restriction imposed above (Section II), and turn to theories which deal with one language in another. In fact, someone who constructed a theory which used English sentences to state their own truth-conditions might, anyway, think of his theory as using his own sentences to state truth-conditions for sentences which another person might use; and in that case the issues are the same. But it is easy to slip into viewing the sort of theory I have been considering from a solipsistic angle, with oneself as the only speaker who needs to be taken into account. That way, some issues are obscured, which become clearer when there is no mistaking the need for interpretation.

In the simplest case, someone interpreting a foreign language will himself already have a name for a suitable object (say a planet): that is, an object such that foreign utterances containing a certain expression can be interpreted as speech-acts about that object, intelligible in terms of propositional attitudes about it, where generation of the content-specifications which represent the speech-acts as being about the object employs clauses like this:

'Aleph' stands for Jupiter.

In a slightly less simple case, he will not antecedently have a name for a suitable object, having had no occasion to use one. But in the course of his attempt to interpret the foreign speakers, his attention might be drawn to some object in their environment, say a mountain (hitherto unknown to him). He would thereby acquire a batch of theory about the mountain. That theory (the facts about the mountain, as he sees them), together with plausible principles about the impact of the environment on propositional attitudes, might make it intelligible that his subjects should have certain propositional attitudes about the mountain, in terms of which he might be able to make sense of utterances by them containing some expression—say 'Afla'—on the hypothesis that it is their name for that mountain.

As far as that goes, he might deal with 'Afla' in his theory of their language by a clause to the following effect:

'Afla' stands for that mountain.

But the context drawn on by the demonstrative would be difficult to keep track of in using the theory. And it might be hard to come by a context-free unique specification to substitute for the demonstrative specification. A neat solution would be simply to adopt the subjects' name for his own use, both in stating the non-semantic facts he has discovered about the mountain and in expressing his theory about their name—thus:

'Afla' stands for Afla.

So far, then, things are much as before. Discussion of two complications should help to make the general picture clearer.

VIII

The first complication is the case of bearerless names, which brings us, I believe, to the deepest source of the richer conception of the sense of names. Suppose an interpreter finds an expression—say 'Mumbo-Jumbo'—which functions, syntactically, like other expressions which he can construe as names, but for which he can find no bearer, and reasonably believes there is no bearer. Such an interpreter, then, can accept no clause of the form

'Mumbo-Jumbo' stands for *b*

where '*b*' is replaced by a name he could use (as above) to express a theory of his own about an object. Names which, in an interpreter's view, have no bearers cannot, by that interpreter, be handled in a theory of sense in the style considered so far. In his view they can have no sense, if a name's having a sense is its being able to be dealt with in that style.

Here we have a genuine divergence from Frege, and one which goes deep. Frege held that the sense of a name, if expressible otherwise than by the name itself, is expressible by a definite description. Definite descriptions are taken to have whatever sense they have independently of whether or not objects answer to them. Thus a name without a bearer could, in Frege's view, have a sense in exactly the same way as a name with a bearer.[22]

The non-Fregean view can be defended on these lines. An interpreter's ascription of propositional attitudes to his subject is in general constrained by the facts (as the interpreter sees them). This is partly because intelligibility, in ascriptions of belief at least, requires conformity to reasonable principles about how beliefs can be acquired under the impact of the environment; and partly because the point of ascribing propositional attitudes is to bring out the reasonableness, from a strategic standpoint constituted by possession of the attitudes, of the subject's dealings with the environment. Now, whether a name has a bearer or not (in an interpreter's view) makes a difference to the way in which the interpreter can use beliefs he can ascribe to the subject in making sense of the subject's behaviour. A sincere assertive utterance of a sentence containing a name with a bearer can be understood as expressing a belief correctly describable as a belief, concerning the bearer, that it satisfies

[22] See above, pp. 28–9; especially (p. 29) the remark that the thought expressed by the sentence 'Odysseus was set ashore at Ithaca while sound asleep' 'remains the same whether "Odysseus" has reference or not'.

some specified condition.[23] If the name has no bearer (in the interpreter's view), he cannot describe any suitably related belief in that transparent style. He can indeed gather, from the utterance, that the subject believes himself to have a belief which could be thus described, and believes himself to be expressing such a belief by his words. That might make the subject's behaviour, in speaking as he does, perfectly intelligible; but in a way quite different from the way in which, in the first kind of case, the belief expressed makes the behaviour intelligible. In the second kind of case, the belief which makes the behaviour intelligible is a (false) second-order belief to the effect that the subject has, and is expressing, a first-order belief correctly describable in the transparent style. This second-order belief is manifested by the subject's action, not expressed by his words. No belief is expressed by his words: they purport to express a belief which could be described in the transparent style, but since no appropriate belief could be thus described, there is no such belief as the belief which they purport to express.[24]

Opposition to this, on behalf of Frege, involves, I believe, a suspect conception of how thought relates to reality, and ultimately a suspect conception of mind. The Fregean view would have to seek its support in the idea that thought relates to objects with an essential indirectness: by way of a blueprint or specification which, if formulated, would be expressed in purely general terms. Whether the object exists or not would then be incidental to the availability of the thought. Underlying that idea is the following line of argument. When we mention an object in describing a thought we are giving only an extrinsic characterization of the thought (since the mention of the object takes us outside the subject's mind); but there must be an intrinsic characterization available (one which does not take us outside the subject's mind), and that characterization would have succeeded in specifying the essential core of the thought even if extra-mental reality had not obliged by containing

[23] That is, describable by way of a transparent, or relational, attribution of belief: see, for instance, W. V. Quine, *Word and Object* (Cambridge, Mass.: MIT Press, 1960), 141–56. Of course that does not stop it being describable by way of an opaque belief-attribution too. Note that there is no withdrawing here from the position outlined in Section IV. From 'He is expressing a belief that Hesperus is visible above the elm tree' (an opaque attribution) one can, according to the principle stated in the text, move to the transparent 'He is expressing a belief, concerning Hesperus, that it is visible above the elm tree'. From there, since Hesperus is Phosphorus, one can get to 'He is expressing a belief, concerning Phosphorus, that it is visible above the elm tree'. But there is no route back from there to the opaque 'He is expressing a belief that Phosphorus is visible above the elm tree'.

[24] Frege's difficulties over truth-value gaps (see Dummett, *Frege*, ch. 12) reflect the pressure he is under, in consequence of his wish to see denoting as a genuine relation between expressions and objects, towards accepting the view of this paragraph.

the object. From this standpoint, the argument for the non-Fregean view outlined above goes wrong in its principle that the thought expressed by a sentence containing a name, if there is any such thought, is correctly describable as a thought, concerning some specified object, that it satisfies some specified condition. That would be a merely extrinsic characterization of the thought expressed; it succeeds in fitting the thought only if reality obliges. If reality does not oblige, that does not show, as the argument suggested, that no thought was expressed after all. For the real content of the thought expressed would need to be given by an intrinsic characterization; and that would specify the content of the thought without mentioning extra-mental objects, and thus in purely general terms.[25]

The conception of mind which underlies this insistence on not mentioning objects, in specifying the essential core of a thought, is the conception beautifully captured in Wittgenstein's remark, 'If God had looked into our minds, he would not have been able to see there whom we were speaking of'.[26] It is profoundly attractive, and profoundly unsatisfactory. Rummaging through the repository of general thoughts which, when we find the remark plausible, we are picturing the mind as being, God would fail to find out, precisely, whom we have in mind. Evidently that (mythical) repository is not the right place to look. God (or anyone) might see whom we have in mind, rather, by—for instance—seeing whom we are looking at as we speak. That sort of thing—seeing relations between a person and bits of the world, not prying into a hidden place whose contents could be just as they are even if there were no world—is (in part) what seeing into a person's mind is.[27]

A proper respect for a person's authority on his own thoughts points, anyway, in a quite different direction. When one sincerely and assertively utters a sentence containing a proper name, one means to be expressing a belief which could be correctly described in the transparent style. One does not mean to be expressing a belief whose availability to be expressed is indifferent to the existence or non-existence of a bearer for the name. The availability of the second sort of belief would be no consolation if the first sort turned out to have been, after all, unavailable.

[25] I owe the formulation in terms of intrinsic and extrinsic characterizations to Brian Loar.

[26] L. Wittgenstein, *Philosophical Investigations* (Oxford: Blackwell, 1953), 217.

[27] Ironically, Wittgenstein's dismantling of the conception of mind deplored here—that which underlies the Fregean idea that the sense of a name is indifferent to the existence of a bearer for it—can be seen as carrying Frege's hostility towards mechanistic psychologism to its extreme (and satisfactory) conclusion.

In practice, an interpreter might say things like 'This man is saying that Mumbo-Jumbo brings thunder', and might explain an utterance which he described that way as expressing the belief that Mumbo-Jumbo brings thunder. That is no real objection. Such an interpreter is simply playing along with his deluded subject—putting things his way. There is no serious reason here for assimilating what he has found out about 'Mumbo-Jumbo' to what, in the case described above (Section VII), he has found out about 'Afla'.[28]

IX

The second complication is the 'Hesperus'–'Phosphorus' puzzle: we can get a clearer view of it by considering it in a context in which the need for interpretation is explicit.

Suppose a smoothly functioning hypothesis has it that the mountain which some of the interpreter's subjects call 'Afla' (Section VII) is also called, by some of them, 'Ateb'.[29] Suppose competence with both names coexists, in at least some cases, with ignorance that there is only one mountain involved. Suppose, as before, that the mountain is new to the interpreter, and that he proposes to take over means for referring to it, in expressing his new theory about it, from his subjects. Since he knows (let us suppose) that there is only one mountain involved, his own needs in geographical description (and so forth) would be met by taking over just one of their names. But a theory of their language which said of both names that, say, they stand for Afla would, as before (Section IV), be incapable of making sense of some utterances. To leave room for the combination of competence and ignorance, an interpreter who follows the strategy of adopting names from his subjects needs, at least in his theory of their language, to use both their names: thus, as before (Section IV), 'Afla' stands for Afla and 'Ateb' for Ateb.

We should distinguish two sorts of case. In the first, the possibility of combining competence and ignorance is an idiosyncratic accident of an individual's language-learning history. Instances of this are provided by people's possession of both 'official' names and nicknames, under which they may be introduced to the same person in situations different enough for him not to acquire knowledge of the truth of the appropriate identity sentence. In this sort of case, it seems plausible that for a speaker

[28] Names of fictional characters are another thing again, not discussed in this paper.

[29] This completes an example of Frege's (from a letter: Dummett, *Frege*, p. 97). The letter is Essay II of this volume.

enlightened about the identity the two names are (aside from stylistic considerations) mere variants of each other (differ merely as objects). A theory of sense which aimed to cover only enlightened speakers, excluding even their dealings with unenlightened audiences, could without trouble use the same name on the right-hand sides of clauses dealing with each of the two. Only an aim of comprehensive coverage—enlightened and unenlightened together—would require a theory to handle the two differently, thereby representing them as differing in sense.

In the second sort of case, exemplified by 'Hesperus' and 'Phosphorus', and, in the most probable filling out of Frege's fable, by 'Afla' and 'Ateb', the difference of sense is more deep-seated. Even if everyone knew that Hesperus is Phosphorus, that would not make the two names differ merely as objects; or, if it did, the names would have changed in sense. If someone used the two names indifferently in talking about Venus, so that we could find no interesting correlation between utterances containing 'Hesperus' and (say) beliefs formed in response to evening appearances of the planet, and between utterances containing 'Phosphorus' and beliefs formed in response to its morning appearances,[30] then he would not be displaying competence in our use of the names—or, rather the fictitious use, corresponding to our use of 'the Evening Star' and 'the Morning Star', which I am considering in order to be able to discuss Frege's problem without raising irrelevant issues about definite descriptions. Such connections, between the use of a name and the sort of situation which prompts the beliefs it helps to express, can be, not merely idiosyncratic facts about individuals, but partly constitutive of a shared language. (This suggests the possibility of a well motivated translation of a pair of radically foreign names—not even culturally related to ours, like 'Hespero' and 'Fosforo'—by our 'Hesperus' and 'Phosphorus'. The strategy of adopting the subjects' names is not essential to the sort of situation we are concerned with.)

The second sort of case can seem to support description theories of the sense of names—at least those which can figure in such examples. But it does no such thing. Recall what, in our terms, the issue would be between the austere conception and treatments of names more congenial to description theories. Would knowledge that 'Afla' stands for Afla and 'Ateb' for Ateb, in the context of suitable further knowledge not directly involving the names, suffice for understanding utterances containing them? Or would one need, rather, knowledge in whose spelling out the

[30] This is meant to be only the crudest sketch of the sort of consideration that counts.

bearer of each name is specified in a more informative way? In the present case, the material from which a description theorist might hope to construct more substantial specifications is the obtaining of more or less systematic differences between the evidential situations which ultimately account for utterances containing the names. And here again it seems clear that to insist on knowledge of those differences is to insist on more than would suffice. A competent speaker need not be reflective about the evidential ancestry of his remarks. They have whatever evidential ancestry they do without his needing to know that they do. And it is their having it, not his knowing that they do, which counts.

It is certainly true that speakers are likely to have opinions about their own propensities to respond to different evidential situations with utterances containing different names. It is our possession of such opinions which confers initial plausibility on the idea that, for instance, the sense of 'visible in the evening' is part of the sense of 'Hesperus'. But we should not let that initial plausibility deceive us. Those opinions are the result of an activity which comes naturally to the self-conscious and theory-seeking creatures we are: namely theorizing, not necessarily with much explicitness, about our own verbal behaviour, as just one of the connected phenomena that constitute our world. In theorizing thus about the place of our speech in our world, we are no better placed than external observers of ourselves; indeed we may be worse placed, if we are less well-informed about the extra-linguistic facts (see Section VIII). Speakers' opinions about their own diverging evidential susceptibilities with respect to names are products of self-observation from, so far as it is accessible, an external standpoint. They are not intimations, from within, of an implicitly known normative theory, a recipe for correct speech, which guides competent linguistic behaviour. It seems that something like the latter picture would have to underlie insistence that, for competence in the use of pairs of names which differ in sense in the deep-seated way, nothing less than knowledge (no doubt tacit or implicit) of the relevant differences in sensitivity to evidence would suffice. The picture is simply a version of the psychologism which Frege rightly rejected (Section V).[31]

Difference in sense between 'Hesperus' and 'Phosphorus' appeared, in the account given earlier (Section IV) of how the austere conception copes with Frege's puzzle, as a reflection of failures of substitution in propositional-attitude contexts. That would not satisfy those who look to

[31] Dummett's discussion of this matter (at p. 123 of 'What is a Theory of Meaning?') seems to me to be vitiated by a non-explicit adherence to the essentially psychologistic idea that mastery of a language is possession of a recipe which guides linguistic behaviour.

the notion of difference in sense (as Frege did) to explain the failures of substitution.[32] The present section yields the following amplification of the picture: the failures of substitution, together with the characteristics which those failures force into a theory of sense, are reflections of two different sorts of underlying situation: first, in the trivial cases, the accidental absence, from some speakers' linguistic repertoires, of propensities to behave in ways construable as evidencing assent to the relevant identity sentences; second, in the deeper-seated cases, the different roles of the names in speakers' more or less systematic propensities to respond to different sorts of situation with different sentences. We can picture the failures of substitution and the differences in sense as, jointly and inseparably, products of our attempts at principled imposition of descriptions in terms of speech-acts, and explanations in terms of propositional attitudes, on to the hard behavioural facts about linguistic and other behaviour, with the point of the imposition being to see how sense can be made of speakers by way of sense being made of their speech. In this picture, the differences in sense are located no deeper than the failures of substitution. I entertain the suspicion that the ultimate source of the desire to see the differences in sense as underlying the failures of substitution, and hence as capable of affording genuine explanations of them, is the psychologism about sense which Frege (officially) renounced.

X

Dissatisfaction with theories which handle names in the way I have recommended is likely to focus on the modesty of any claims we could make on their behalf. The sense of a name is displayed, in these theories, by the deductive powers of a clause intelligible only to someone already competent in the use of the very name in question, or else another name with the same sense. This exemplifies, for the case of names, one strand of the quasi-technical notion of modesty, in a theory of meaning, discussed by Dummett in 'What is a Theory of Meaning?'.[33] Dummett makes it appear that a theory which was ineradicably modest, in that sense, would amount to a repudiation of the concept of sense.[34]

[32] Simon Blackburn made me see this.

[33] Dummett's formulation in terms of concepts (p. 101) may suggest that only general terms are in question, but that is not his intention.

[34] In the Appendix, which (among other things) relates the arguments of the lecture (not originally couched in Fregean terminology) to Frege: see esp. pp. 126–8.

In this section I want to sketch the reason why I think his argument unconvincing.[35]

As noted earlier (Section V), Dummett talks of a theory of understanding as specifying (implicit) knowledge in whose possession understanding actually consists; but we can let his argument begin, at least, without jibbing at that. He insists that it is worthless to consider a practical capacity on the model of knowledge of the members of an articulated set of propositions, if one can give no account of what it would be to know the individual propositions. One thing such an account must do is distinguish knowledge of one of the relevant propositions from mere knowledge that some sentence (which in fact expresses it) expresses a truth; the latter state would not be to the point. Now in section VI above some simple remarks seemed adequate to ensure, in the case of the knowledge that 'Hesperus' stands for Hesperus which I claimed would suffice, in the context of other knowledge not directly involving the name, for understanding the name, that it did not crumble into mere knowledge that the sentence '"Hesperus" stands for Hesperus' expresses a truth. Why not say, then, that those remarks constitute (for that case) something meeting Dummett's demand that we specify not merely what would be known in the hypothetical state of knowledge, but also what knowing it would consist in? What is counted as constituting a manifestation in linguistic behaviour of the hypothetical state of knowledge (or, better, of the capacity with the name for which it is claimed that that knowledge, in its context, would suffice) is, according to that suggestion, something like this: whatever behaviour would manifest an ability to use the name, or respond intelligently to uses of it, in speech-acts construable as being about the planet; more specifically, speech-acts in stating whose content we can use the name which appears on the right-hand side of the relevant clause, in a way appropriately tied to the occurrence, in the utterance being interpreted, of the name mentioned on the left-hand side of the relevant clause.

If we give that sort of description, or purported description, of a component of the articulated practical capacity which, according to a theory of a language, constitutes mastery of it, we make essential appeal to the interpretation of the language afforded by the theory itself. The manoeuvre preserves modesty, in the sense outlined above; no one could employ the suggested account, in order to ascertain from someone's behaviour whether he had the relevant ability, without being able to

[35] I regard what follows as a minimal defence of my continuing to attempt to locate myself in a place which Dummett thinks he has shown non-existent. His subtle and powerful argument needs much more discussion.

understand sentences like 'This man is engaging in behaviour construable as his saying that Hesperus is visible above the elm tree'—that is, without himself already being able to understand 'Hesperus'.

Now Dummett's exposition of the notion of modesty amalgamates two notions: first, the notion of a theory which, as above, refuses to make itself intelligible except to someone who already understands the expressions it deals with (or others with the same sense); second, the notion of a theory which refuses to say what would count as a manifestation of the individual component abilities into which it purports to segment the ability to speak the language—that is, refuses to say not only what would be known, in the knowledge of a structured theory which, it is claimed, would suffice for understanding the language, but also what the ability corresponding to each piece of this hypothetical knowledge would consist in.[36] Thus, on pain of the notion of modesty falling apart, those purported specifications of component abilities suggested above, since they preserve modesty in the sense of the first component notion, must not be allowed to count as other than modest in the sense of the second. They are debarred, then, from being accepted as effecting a genuine segmentation of the ability to speak a language. Dummett equates refusal to say what the knowledge in question would consist in, as well as what would be known—modesty in the sense of the second component notion—with repudiation of the concept of sense.[37] Thus the amalgamation reflects the position mentioned at the beginning of this section: that is, that a theory which is ineradicably modest in the sense of the first component notion cannot but repudiate the concept of sense.

Why are those theory-presupposing specifications not allowed to count? Since they are not even contemplated, Dummett's lecture contains no explicit answer. It is easy to guess, however, that they do not occur to him, even as candidates to be ruled out, because he assumes that the acceptability of the demand to say what the knowledge would consist in is the acceptability of a demand for a reduction. The idea is that a genuine segmentation of the ability to speak a language would segment it into component abilities describable, as nearly as possible, in purely behavioural terms. And the theory-presupposing specifications refuse to attempt any such reduction.[38]

[36] Note, in 'What is a Theory of Meaning?' on p. 101, the oscillation between explanation of concepts and explanation of what it is to have concepts.

[37] See n. 34.

[38] That this is how the demand for segmentation is understood emerges, I believe, from Dummett's employment of it to make difficulties for realist (truth-conditions) theories of

Why, though, should reduction be either necessary or desirable? It is extraordinary that a reductive construal of the demand that one say what the states of knowledge in question would consist in should seem so obvious as not to need explicit acknowledgement, let alone defence.

Dummett's lecture is shot through with intimations of an idea which would make that intelligible: namely the idea that a theory of a language ought to be such that we could picture implicit knowledge of it as guiding competent linguistic behaviour.[39] Obviously a theory could not perform that service if understanding the theory required an exercise of the very ability it was to guide. If a theory is to guide speech it must generate instructions for doing things with sentences, intelligible independently of understanding those sentences. That would make congenial the search for a theory which—to illustrate with a simple case—might, in the case of a sentence combining a name with an unstructured predicate, generate instructions on this pattern: 'First find the object (if any) which is thus and so. Then apply such and such tests to it. If the outcome is thus and so, adopt a preparedness to volunteer the sentence or utter "Yes" on hearing it.'[40] A theory which systematically generated such instructions, on the basis of structure within sentences, would presumably be compelled to handle names by means of clauses which specified ways of identifying or recognizing objects; thus richness (in the sense of Section V), as opposed to austerity, is the shape which immodesty takes in the case of names.

If pressed to its extreme, the idea from which that line of thought begins—the idea that a theory of a language ought to be such that we could picture implicit knowledge of it as guiding competent linguistic behaviour—leads to the incoherent notion of a theory which cannot be stated in any language.[41] Coherence might be preserved by demanding no more than an approximation, as close as may be, to the impossible ideal of a theory we could state to a person in order to teach him to talk. But in any case the idea which underlies these aspirations—the idea that

meaning: see 'What is a Theory of Meaning? (II)', *passim*, and cf. *Frege*, 467. 'As nearly as possible' is meant to allow, as Dummett does, for the employment of concepts like that of assent (see 'What is a Theory of Meaning? (II)', 80).

[39] He speaks throughout of theories of meaning as being *used to obtain* understanding of a language (e.g. p. 114). Note also his talk of 'the undoubted fact that a process of derivation of some kind is involved in the understanding of a sentence' (p. 112). See also n. 31 above.

[40] Another part of the theory of a language (the theory of force) is needed to get one from this sort of thing to an ability actually to engage in conversation: see 'What is a Theory of Meaning? (II)', e.g. at pp. 72–4. It is clear that such a theory, on this conception, will have a great deal of work to do.

[41] See 'What is a Theory of Meaning?', 103–4, where Dummett comes close to imposing the incoherent requirement on those full-blooded theories of meaning which he recommends.

linguistic behaviour is guided by implicit knowledge—is nothing but a version of the psychologism which Frege denounced and which Dummett officially disclaims.

If we try to preserve everything Frege said about sense, we characterize a position with an internal tension. On the one hand, there is Frege's anti-psychologism, which in Wittgenstein's hands transforms itself into a coherent and satisfying view of the mind's place in reality, stably intermediate between the crass extremes of behaviourism and a psychologism which is objectionable not because it is mentalistic but because it is pseudo-scientific. On the other hand, there is the idea that a theory of sense would be rich or immodest. My claim is that the latter idea can only be justified on the basis of vestiges of the psychologism rejected in the first component. If I am right, something has to be repudiated. It is a terminological issue whether 'sense' belongs with the second component; but fairness to Frege seems to justify trying to find a place for his terminology in a position purged of those elements which can only be grounded in hidden psychologism. Modesty in our demands on a theory of meaning is, in this view, not a repudiation of Frege's notion but an insistence on the feature which makes it an effort at something truly great; immodesty is not a vindication of Frege but a betrayal.[42]

XI

An adherent of a causal theory of names need not, in the interests of his theory, be unsympathetic to what I have said so far. His concern would be, not to enrich the right-hand sides of those specifications of denotation which I have been considering, but to say something substantial about the relation.[43]

Deduction within a truth-theory moves from axioms which assign semantic properties to sentence-constituents (for instance, denoting some specified thing), by way of clauses dealing with modes of combination, to theorems which assign truth-conditions to sentences. That is what warrants the claim that such a theory displays the sense of a sentence-constituent as, in Frege's metaphor, its contribution to the truth-conditions of sentences in which it occurs. The deductive direction

[42] I hope it is clear that I want no truck with the theory devised by Dummett in his Appendix under the title of 'holism'. That theory is an attempt (wholly unconvincing, as Dummett rightly says) to meet the reductionist demand which I reject. The topic of holism is too difficult (and confused) to be dealt with *ambulando* in this paper.

[43] *Pace* (apparently) Dummett, 'What is a Theory of Meaning?', 125.

can make it seem that the whole structure floats unsupported unless the nature of the semantic properties and relations from which the derivations start is independently explained. Thus it can seem that we need a general account of what it is for a name to denote something, conceptually prior to the truth-theory itself, from which the truth-theory's assignments of denotation to names can be seen as derivable, and in terms of which, consequently, they can be seen as explained. A causal analysis of the relation between a name and its bearer might seem well suited to meet that apparent need.[44]

But the need is only apparent. It is not true that we condemn a truth-theory to floating in a void, if we reject the alleged obligation to fasten it directly to the causal realities of language-use by way of its axioms. On my account, those truth-theories that can serve as theories of sense are already anchored to the facts of language-use at the level of their theorems: the anchoring being effected by the requirement that assignments of truth-conditions are to be usable in specifications of content of intelligible speech-acts. Since the theorems must be derivable within the theory, the requirement bears indirectly on the theory's deductive apparatus, including assignments of denotation; and the deductive apparatus needs no attachment to the extra-theoretical facts over and above what that affords. Thus we can acquire such understanding as we need of the deductive apparatus (in particular, of the denoting relation) by reversing the theory's deductive direction, starting from our understanding of the requirement of serviceability in interpretation imposed on its consequences. We grasp what it is for a name to denote something by grasping the role played by the statement that it does in derivations of acceptable assignments of truth-conditions to sentences—assignments, that is, which would pull their weight in making sense of speakers of the language we are concerned with.[45]

According to Frege, it is only in the context of a sentence that a word has meaning.[46] What he meant was that we should not look for accounts of the meaning of particular words except in terms of their contributions to the meanings of sentences. But it seems in the spirit of his slogan to suggest also, as above, that we should not look for accounts of the sorts of meaning possessed by words of general kinds—for instance, an account of denotation, as possessed by names—except in terms of

[44] Cf. Hartry Field, 'Tarski's Theory of Truth', *Journal of Philosophy*, 69 (1972), 347.

[45] Cf. Davidson, 'In Defense of Convention T', in H. Leblanc (ed.), *Truth, Syntax and Modality* (Amsterdam: North-Holland, 1973), esp. p. 84.

[46] G. Frege, *The Foundations of Arithmetic*, trans. J. L. Austin (Oxford: Blackwell, 1953), 73.

the contributions made by words of those kinds to the meanings of sentences.

To reject the quest for a causal analysis of denotation, conceived as prior to interpreting a language by constructing a truth-theory, is not to deny the relevance of causal relations in determining what a name denotes. In fact causal relations will be involved in any adequate elaboration of the requirement imposed on a theory's consequences. To illustrate: suppose a candidate theory of a language would have us describe a certain speaker as saying that *p*, where we imagine '*p*' replaced by a sentence which mentions a particular concrete object. Can we make sense of his saying that *p*? Standardly, we make sense of sayings as expressing the corresponding belief. And ascription of the belief that *p* to our speaker is constrained by a principle on the following lines: one cannot intelligibly regard a person as having a belief about a particular concrete object if one cannot see him as having been exposed to the causal influence of that object, in ways suitable for the acquisition of information (or misinformation) about it.[47] Such principles, operating in that ascription of propositional attitudes which we need to go in for in order to make sense of linguistic behaviour, make causation crucial in the determination of what at least some names denote.

There is, however, not the slightest reason to expect that one could construct, out of such materials, a general relational formula true of every name and its bearer. And even if a formula with the right extension could be constructed out of these materials, it would not constitute that prior fixed point of suspension for a truth-theory which was dreamed of in the argument sketched at the beginning of this section. The ultimate justification for an assignment of denotation would be, not some causal relation between an object and utterances of the name, accessible independently of interpreting the language, but—as ever—the acceptability of interpretations which that assignment helps to confer on whole sentences.

Saul Kripke, who is often described as a proponent of a causal theory of denotation for names, in fact expressed the suspicion that any substantial theory of names—like any philosophical theory—is most likely to be wrong.[48] I think Kripke's suspicion was well placed. In this essay I hope to have indicated how we might find that situation possible to live with.

[47] See Gareth Evans, 'The Causal Theory of Names', *Aristotelian Society*, Supp. vol. 47 (1973), now Essay XII of this volume, at pp. 217–20 below.

[48] See 'Naming and Necessity', in Donald Davidson and Gilbert Harman (eds.), *Semantics of Natural Language* (Dordrecht: Reidel, 1972), 280, 300–3.

VIII

WHAT DOES THE APPEAL TO USE DO FOR THE THEORY OF MEANING?

MICHAEL DUMMETT

Consider the following style of argument. What would one say, e.g., 'Either he *is* your brother or he *isn't*', *for*? Well, it is tantamount to saying, 'There must be a definite answer: there are no two ways about it.' We say this when someone is shilly-shallying, behaving as if it were no more right to say the one thing than the other: so the utterance of that instance of the law of excluded middle is an expression of the conviction that the sentence, 'He is your brother', has a *definite* sense. That, therefore, is the meaning of the sentence, 'Either he is your brother or he isn't': that is its *use* in the language.

No doubt everyone here would agree that that is a bad argument: but why is it a bad argument? A superficial answer might be, 'It does not take account of other uses that exist for uttering an instance of the law of excluded middle, for example in the course of a deductive argument. Thus Littlewood proved a theorem by showing that it followed both from the Riemann hypothesis and from the negation of that hypothesis: so his proof might have started, "Either the Riemann hypothesis is true or it is false."' This is a superficial answer, because, although it is quite true that people do use instances of the law of excluded middle in this way, they might, given classical logic, perfectly well not do so, and still be able to carry out all the deductive arguments that they wanted to; and yet the philosophical argument with which I started would still be a bad argument. The following explanation of this fact is a great improvement. The recognition of the law of excluded middle as valid hangs together with the admission of certain forms of inference as valid, in particular, the dilemma or argument by cases:

$$\frac{\text{If A, then B} \qquad \text{If not A, then B}}{\text{Therefore, B}}$$

From *Meaning and Use*, ed. A. Margalit (Dordrecht: Reidel, 1979), 123–35. Reprinted by permission of Kluwer Academic Publishers.

which underlies the proof of Littlewood's that was mentioned. It hangs together with it in the sense that any reasonable general formulation of these rules of inference, together with a few others that strike us as inescapable, will result in our being able to deduce each sentence of the form 'A or not A' from no hypotheses at all (for instance, by an argument whose last step is the dilemma, as above, with 'B' replaced by 'A or not A'). The notion of truth is, of course, connected with that of a valid inference by the fact that whatever follows by valid inferences from true premisses must be true: so we are committed, if we accept the dilemma and related forms of argument, to regarding a sentence 'A or not A' as true. Now the meaning of a sentence is more closely connected with what, if anything, does or would render it true than with what would prompt an actual utterance of it. Hence, an understanding of a sentence of that form is to be sought by explaining those meanings of the logical constants 'or' and 'not' which permit of its derivation from the null set of hypotheses.

The second argument does not, like the first, tamely accept that the use of a sentence, in the sense of the point that an utterance of it might have, determines its meaning, and then claim that some such uses have been overlooked. Rather, it challenges that principle by giving reasons for thinking that we must have a prior understanding of the *sentence* before we can be in a position to ask what the point of a particular utterance of it may be. The argument, as stated, appeals to an already understood notion of truth, with a known connection with our recognition of any given principle of inference. The proponent of the argument that is being criticized may feel that such a notion of truth is spurious, and has available a well-known device for countering an appeal to it: he declares that the whole explanation of 'true', in its only intelligible sense, is given by the principle that 'A' is equivalent to 'It is true that A', or by a definition that is just sufficient to yield this equivalence for each case; the use, or meaning, of an assertion that a sentence is true will then be precisely the same as that of an utterance of that sentence, and the notion of truth will be impotent to yield any results about meaning not previously obtained by inquiring into use. But the sense of 'true' required for the second counter-argument is shown by that argument itself: what is needed of a true sentence is that there should exist means of justifying an assertion of it of a kind we are accustomed to accept elsewhere; and so the word 'true' can be dropped from the argument, and a direct appeal made to this notion. This is, indeed, to assume that we recognize certain general principles for the justification of our assertions; but so we obviously do, otherwise there would be no such thing as deductive argument. It is now open to the proponent of the second counter-argument to concede that, given classical logic, an instance of the law of excluded middle is so obvious that the point of an assertion of it is scarcely ever to call attention to the fact that it can be justified, without calling in question his own thesis that it is the possibility of justification to which our primary understanding of the sentence relates; indeed, he can even maintain that 'A or not A' would be true in some, or all, cases in which the sense of 'A' is *not* definite, so that an assertion of it, intended to have the point which the proponent of the original argument rightly said that such assertions often have, would go awry otherwise than by failing to be true.

Now I do not for one moment suggest that an argument of the style with which we started out and with which some of us were made, during a period now passed, wearisomely familiar, represents faithfully the notion of use which Wittgenstein had in mind when he coined the slogan, 'Meaning is use.' On the contrary, Wittgenstein's notion was a much more general one: it comprised anything that could be counted as belonging to the role of the sentence in the language-game, which certainly included, not only the communicatory function of an utterance of the sentence itself, but also that of an utterance of a complex sentence of which it was a constituent and, as well, such other features as were appealed to in the counter-argument I set out. To say this, however, is to make the conception of meaning as use totally programmatic: any feature of our linguistic practice that relates to the sentence may be cited as bearing on its meaning. There is, however, a reason why Wittgenstein's later philosophy of language should have led to this misapplication of the identification of meaning with use. This lies in Wittgenstein's repudiation of the Fregean distinction between sense and force and, particularly, of Frege's idea that there is such a thing as assertoric force in general. There are, on this conception of Frege's, three grades in understanding an assertoric utterance. First comes the grasp we have of the sense of the sentence, of the thought expressed; and this consists in an understanding, which is derived in accordance with our apprehension of the construction of the sentence out of its linguistic elements (let us say, inexactly, its component words), of the condition which must obtain for the sentence to be true. Secondly, there is our knowledge of the practice of assertion: the speaker is not merely uttering a sentence with which is associated a certain truth-condition, that is, expressing a certain thought, but saying (that is, asserting) that that thought is true, as opposed to asking whether it is true, supposing it to be true for the purposes of argument, declaring himself unwilling to deny that it is true, advising his hearer to make it true, expressing the wish that it were true, or the like. (Whether or not there is a non-circular account of what it is to assert that a thought is true, that is, of what is effected by an assertoric utterance of a sentence expressing that thought, is another matter.) And, finally, there is the divination of the speaker's particular intention in asserting that thought to be true on that particular occasion. Wittgenstein rejected this conception, on the ground that there is no such thing as 'the practice of assertion', or as, in his terminology, the language-game of assertion, considered as effected by the utterance, in assertoric mode, of any sentence syntactically fitted to be used assertorically and to which we may ascribe conditions for it (or a specific utterance of it) to be true or false.

> Something happens—and then I make a noise. What for? Presumably in order to tell what happens.—But how is *telling* done? When are we said to *tell* anything?—What is the language-game of telling?—I should like to say: you regard it much too much as a matter of course that one can tell anything to anyone. That is to say: we are so much accustomed to communication [*Mitteilung*—the abstract noun cognate with the verb used for 'to tell'] through language, in conversation, that it looks to us as if the whole point of communication lay in this: someone else grasps the sense of my words—which is something mental: he as it were takes it into his own mind. If he then does something further with it as well, that is no part of the immediate purpose of language.[1]

Now the fact is that it is difficult to obliterate the distinction between the first grade of understanding and the second without thereby also obliterating that between the second and the third. This is because our concept of truth gets a large part of its point from the contrast that we wish to draw between a statement's being true and any more primitive, or at least undifferentiated, conception of its being appropriate: for instance, between its being true and the speaker's having a sufficient warrant to take it as true, or between its being true and the intention that the speaker had in asserting it to be true just then being a just one, his having had a legitimate point in making it. Of course, once we have any given conception of a particular sentence's being determined, in some objective manner, as true or as false, then these distinctions arise naturally, indeed inevitably: the questions of interest are why we introduce the notion of truth at all, and why, in doing so, we draw the line between the condition for a statement to be true and the condition for a speaker's being in the right in making it in these more general ways at just the place we do, and not somewhere else. There are various

[1] *Philosophical Investigations*, trans. G. E. M. Anscombe (Oxford: Blackwell, 1974), i. 363.

correct answers to these questions, one being the necessity of explaining the role of the sentence when it figures as a constituent in more complex sentences; but this is not our present concern. Another partial answer is, obviously, the dependence of a speaker's point in making a statement on the context, something which, if we are to attain a conception of the meaning of the sentence as a type, we must either filter out or reduce to a definite rule (as we can explain indexical expressions systematically). But this is not an important feature of the objection to the account with which I started of sentences like 'Either he is your brother or he isn't', an utterance of which was claimed as only ever having one kind of point; as I remarked, the account would be wrong even if the claim were sound. Rather, in that case, what we appealed to was the existence in the practice of the speakers of certain generally accepted procedures for justifying statements, procedures which would always yield a justification of an instance of the law of excluded middle even if it was never in fact invoked in such a case. This looks circular, since such procedures are for the purpose of justifying a statement *as true*, rather than as making a sound point. So it comes down to this: that our linguistic practice—the language-games in which we participate—involves the process whereby those utterances which we call assertions (and perhaps some others) are subject to challenge by our hearers and the process of responding to such challenges; and, if we were to try to give any account of these practices, a mastery of which is certainly essential for the ability to engage in converse with others, we shall be forced to distinguish between different types of such challenges, according to the kind of response that is appropriate; and among these are challenges as to the truth of what is said and challenges as to its point (the latter of which doubtless further subdivide into challenges as to relevance, as to implicature in Grice's sense, etc.). Here a challenge as to truth is to be distinguished by the fact that, if successfully met, the challenger will himself give assent to the statement (though he need not be prepared himself to make that statement, since it may be objectionable in other ways, e.g. as breaking a confidence or being insulting); hence Quine's properly placed emphasis upon the notions of assent and dissent. (A suspicion of circularity arises here also, since an expression of assent is surely an expression of a willingness to make the statement so far as its truth is concerned, that is, but for possible objections to it which are not objections as to truth; but I will not push the inquiry further. Of course, as I said, these distinctions are easy to draw once we have the notion of truth and know its application to a given sentence; but I have been concerned with what we need the notion for and why we give it the application that we do.) This is not to say that the notion of truth so arrived at will serve all the purposes for which we need it, for instance to explain the behaviour of a sentence when it is a constituent of a more complex sentence.

Once we have the notion of truth and so can distinguish between the second and the third grade of understanding, the distinction between the first and second grade is all but inevitable. If all utterances were assertoric, and no sentence ever occurred as a constituent of another sentence, there would indeed be no place for it, but this would make no difference to the present argument: so long as we appealed to the notion of the truth-conditions of a sentence as determining its particular content, we should, in explaining what is effected by an utterance of a sentence, have to give a general description of the linguistic practice of making assertions. But the notion of truth is precisely what we need, or, rather, what is forced on us, if we wish to distinguish between the second and third grades of understanding. Hence, if there is no such thing as the general practice of making assertions—or, at least, as a uniform description of what this practice consists in, for a sentence with an arbitrarily given individual content—then there can be no distinction, at any rate no general distinction, between the second and third grades of understanding either. And what this appears to mean is that any account of the meaning of a given sentence must simultaneously explain every feature of the significance of any possible utterance of it. This, indeed, is not particularly difficult to do for any one sentence, at least if we ignore utterances the point of which is heavily context-dependent. What seems impossibly hard is to construct a systematic theory of meaning for a language along these lines, that is, one which would show the derivation of the significance of the sentence in accordance with its composition: as soon as we begin to think about the construction of such a theory, we at once start to segment the task it has to accomplish, along the lines of the repudiated distinctions between truth-conditions, force and point. Wittgenstein's repudiation of these distinctions is expressed by his adherence to the redundancy theory of truth (expressed in a characteristically sloppy manner in the *Remarks on the Foundations of Mathematics*[2]—"For what does a proposition's '*being true*' mean? '*p*' *is true* = *p*. (That is the answer.)"). If the equivalence of ' "Snow is white" is true' with 'Snow is white', and so on, constitutes the *whole* explanation of the concept of truth, then the concept is useless in giving a theory of meaning. It is because of his rejection of the concepts of assertion and of

[2] Ed. G. H. von Wright, R. Rhees and G. E. M. Anscombe and trans. G. E. M. Anscombe (Oxford: Blackwell, 1978), i. App. I: 6.

truth as capable of playing any role in an account of how language functions that Wittgenstein's identification of meaning with use lent itself to the kind of misapplication with which I began.

In an earlier period, however, Wittgenstein had seized on the notion of the justification of a statement as the key to an explanation of sense: "It is what is regarded as the justification of an assertion that constitutes the sense of the assertion."[3]

It is natural to contrast the idea that meaning is given by determining what justifies us in using a sentence to make an assertion with the idea that it is given by determining the conditions under which the sentence is true; and Wittgenstein certainly meant to convey a very sharp divergence from the theory of meaning in terms of truth-conditions set out in the *Tractatus*. But to say that, on the view of meaning Wittgenstein held in the intermediate period, the meaning of a sentence is *not* determined by its truth-conditions, is liable to misconstruction. If one holds that the meaning of a sentence is given in terms of what has to hold for it to be true, it is open to one to say that someone may understand a sentence although he has not yet learned by what means we may recognize it as true, nor, therefore, what justifies an assertion of it. But there is an asymmetry here: thinking that the meaning of a sentence is given by what justifies asserting it does not entitle one to suggest that someone might understand a sentence without yet knowing the condition that must obtain for it to be true. Rather, from such a standpoint one would say that the only legitimate notion of truth is one that is to be explained in terms of what justifies an assertion: a sentence is true if an assertion made by means of it would be justified. (Or, possibly, if there is some recognizable state of affairs such that, if a speaker knew of it, he would be justified in making that assertion.) There is therefore a sense in which, even on the theory of meaning which is opposed to that of the *Tractatus*, it remains the case that the sense of a sentence is determined by its truth-conditions: the question is what is the relation between the notion of truth and that of the justification of an assertion.

The theory of meaning expressed by the remark I have quoted from the *Philosophical Grammar* stands in opposition to any conception of meaning under which the sense of a sentence is given in terms of a notion of truth taken as objectively and determinately either attaching or not attaching to each sentence independently of our knowledge, or capacity to know, whether or not that sentence is true, a notion of truth

[3] *Philosophical Grammar*, ed. R. Rhees and trans. A. P. Kenny (Oxford: Blackwell, 1974), i. 40.

which is therefore taken to be grasped without reference, in all cases, to the means available to us for judging a sentence to be true. This conception plainly informs the *Tractatus*. That work carries a fundamental commitment to the principle of bivalence, integral to the conception of meaning I have just characterized, since, if bivalence did not hold, the truth-tables would not have the kind of importance allotted to them in the *Tractatus*. Since the *Tractatus* also contends that our understanding of our sentences involves the grasp of infinitary truth-operations, it is equally plain that a grasp of the meaning of a sentence is not held to be in all cases related to, or given in terms of, the means available to us for recognizing it as true. Wittgenstein came to repudiate this conception for a multitude of reasons. First, a theory of meaning of this kind is powerless to explain how we come by our knowledge of the conditions which warrant us in asserting a statement, the means by which we can establish a statement as true. Granted that the meaning of a sentence is not, in the first place, given in terms of how we recognize it as true, still our grasp of what is to count for us as showing that it is true must be *derived* in some way from our knowledge of its meaning: for, if not, then, even when the meaning of a sentence has been fixed by determining its truth-conditions, there will still remain room for decision as to what we shall choose to count as showing it to be true, and this is counter-intuitive. But, once we have allowed the two notions, that of truth and that of the means by which truth is recognized, to be sundered at the outset, we shall never find a means to connect them up again, to explain how the one is derived from the other. More generally, the theory violates the intuitive connection between meaning and knowledge: two sentences may, according to the *Tractatus* theory, express the same sense (because they make the same division in logical space) without our perceiving that their senses are the same; this is because sense has been thought of as given in terms of what *makes* a sentence true, rather than in terms of how it is recognized as true. But, on the contrary, we ought to say that the meaning of any expression is determined by what a speaker must know if he is to be said to understand that expression; it follows that, if someone understands two expressions that have the same meaning, he must know that their meaning is the same. (It could be argued—and has been argued by me—that this is in part a consequence of Wittgenstein's abandonment of the distinction between sense and reference as drawn by Frege, and therefore not a necessary result of holding a conception of meaning as given by truth-conditions, under a notion of truth subject to bivalence and not directly connected, in all cases, with our means of recognizing truth. That is not to say that the more restricted form of the objection, that such a conception of meaning will not always allow an explanation of how we derive the means of recognizing the truth of a sentence from the condition for it to be true, is met by appealing to a Fregean distinction between sense and reference.)

Secondly, the *Tractatus* conception leaves us unable to state informatively the conditions for the truth of many of our sentences: an essential circularity appears in any attempt to do this, a circularity which does not appear in a characterization of what *justifies* us in asserting a sentence. This circularity then leads us to attribute to a speaker a capacity for immediate *recognition* of certain qualities, objects, processes or states (the private ostensive definition), which capacity can be no further explained. But now it is evident that this attribution is idle: all would go on in just the same way if the speaker misrecognized the entity every time, or if there were nothing there to be recognized; at least, it would do so provided that, whenever recognition by two or more speakers was called for, they tended to make the same mistakes. The conception of our apprehension of the truth-condition of a sentence, with its attendant capacity for immediate recognition of the presence of the referents of certain terms, therefore fails to be explanatory: what, ultimately, actually justifies us in the assertions we make is the fact of agreement between speakers, which need not be taken as resting on anything more basic. As for the notion of truth, the circularity disappears when we cease to think of it, as the *Tractatus* insists, as that in terms of which the meanings of our sentences are given, and, instead, regard its application to a sentence as explicable only after the sense of that sentence is known.

Thirdly, this same circularity attends our attempts to state the truth-conditions of those sentences our alleged grasp of which transcends our means of recognizing them as true: our grasp of the condition for such a sentence to be true cannot consist in our ability, in certain special cases, to recognize it as true, just because it involves our awareness that it may still be true even when we are unable to recognize it as such. But then to attribute to us a grasp of the condition for a sentence to be true, under such a transcendental notion of truth, violates the principle that meaning is use: for a knowledge of the condition for the truth of the sentence of this kind cannot be fully manifested by the use the speaker makes of it, that is, by the linguistic and non-linguistic behaviour on his part that is connected with the utterance of the sentence.

These arguments are all of a negative kind: in so far as they are cogent, they show the conception of meaning as given by truth-conditions, as found in Frege, and, in a different form, in the *Tractatus*, to be inadmissible; but they do not show the conception of it as given in terms of what justifies an assertion, rather than of some other feature of the use of a sentence, to be correct. In so far as they are merely negative arguments, they survive into Wittgenstein's later period: but I have suggested that he came to adopt a still more radical view, one involving a repudiation of the sense/force distinction in a way in which the idea of meaning as given in terms of the justification of an assertion does not. (Indeed, it is of importance that the formulation I quoted from the *Philosophical Grammar* employs the notion of assertion.) I cannot here attempt an evaluation of the negative arguments I have just sketched. Instead, I want to argue that the thesis that what constitutes a justification of an assertion determines, or shows, the sense of the sentence asserted indicates very precisely the constraints which the identification of meaning with use puts, and those it does not put, upon a theory of meaning.

First, the constraints it does not impose. If someone has the idea that the justification for an assertion is the key to its content, he is allowing himself a much richer set of data to which to appeal than those which either Quine or Davidson permits himself. For Quine, the relevant data consist solely of the correlations between the sensory stimuli to which a speaker is subjected and his readiness to assent to or dissent from a sentence. Davidson is more generous in allowing correlations between a speaker's holding a sentence true and prevailing conditions of any kind. Both, however, propose to construct a translation manual or meaning-theory by appeal solely to data of the form of answers to the question, 'When do the speakers hold, or acknowledge, sentences as true or as false?' The reason for this limitation is made very explicit by Quine, less so by Davidson. For both, the problem is of constructing a translation-manual, or meaning-theory, for a language whose speakers one may observe and perhaps even interact with, but which is previously quite unknown. All that one has to go on is what one can see and hear of the speakers' utterances and associated behaviour. Now perhaps one can make a plausible identification, in behavioural terms, of the speakers' mode of expressing assent and dissent. But to go any further would be illegitimate. The notion of justification, as used by Wittgenstein, does not refer only to the process of justifying an assertion, when challenged, as this takes place within the language, since various things are involved in a full account of what justifies certain assertions that the speakers tacitly take for granted and would never explicitly cite. Nevertheless, it does involve all that would be appealed to in justifying an assertion, as this occurs between speakers. But, from the standpoint of Quine and

Davidson, if I have understood them aright, to appeal to such a complicated thing as the justifications which speakers give of their assertions is out of the question, since, to become aware of that, or even to recognize what constituted a demand for justification and what a response to it, one would have already to understand a large part of the language.

Philosophers, unlike historians, do not have to solve problems that are clearly demarcated in advance; and so they make up their own problems—set themselves tasks, and then try to perform them. Disputes over philosophical methodology are largely about which are the right problems to set. One can hardly prove that this or that is the right problem: that would be possible only if, behind the problems philosophers try to solve, lay further clearly defined problems, and the solution to the former were a means to the solution of the latter. The question is only the vague one: by solving which problems shall we gain philosophical illumination? Now what is the point of posing the problem: how we should arrive at an interpretation of a language hitherto quite unknown to us? It is, surely, to exclude from the description of the interpretation or of the process of arriving at it any appeals to concepts which covertly presuppose an understanding of the language. But the consequence of so posing the problem is that we fasten on some feature of the speakers' linguistic behaviour which can be described at the outset, before any understanding of the language has been gained, and try to use it as the basis for the entire interpretation. Language is, however, an enormously complicated thing, and it is highly unlikely that a satisfactory interpretation of it is accessible if we so restrict ourselves. Certainly our actual acquisition of our mother-tongue proceeds by stages: some features of our linguistic practice can be mastered only after others have already been mastered. We are, at any rate, not now in the position of having to interpret some radically foreign language: that is a practical problem, the solution to which is not obviously necessary or sufficient for the kind of understanding of how language functions which we, as philosophers, wish to attain. We already have, in our language, expressions for various concepts which relate to our use of language, among them that of the justification of an assertion. What we want to arrive at is a model of that in which our understanding of our language consists, a model which will be adequate to explain the entire practice of speaking the language. Certainly that model must itself be described in terms which do not presuppose a tacit understanding of terms, such as 'assertion', 'justification', 'true', etc., which relate to the practice of which the model aims to provide an account, or it will, to that extent, fail to be explanatory. But that does not mean that, in groping our way towards such a model, we must eschew appeal to any of those concepts which are not to be used in giving the model itself. It does not matter whether or not an outside observer—a Martian, say, who communicated by means so different from our own that he would not for a long time recognize human language as a medium of communication—could ever arrive at the model we hope to give: all that matters is whether, once he had it, it would serve to make our language intelligible to him.

Now for the constraints which the identification of meaning with use does impose. Having expressed a point of basic methodological disagreement with Quine, let me now record one of strong agreement with him. In his lecture 'Mind and Verbal Dispositions',[4] Quine says, "when I define the understanding of a sentence as knowledge of its truth conditions I am certainly not offering a definition to rest with; my term 'knowledge' is as poor a resting-point as the term 'understanding' itself," and goes on to ask, "In what behavioural disposition . . . does a man's knowledge of the truth conditions of the sentence . . . consist?" This is in full consonance with what I have myself repeatedly insisted on, that a meaning-theory, being a theoretical representation of a practical ability, must not only say *what* a speaker must know in order to know the language, but in what his having that knowledge consists, that is, what constitutes a manifestation of it. (I disagree with Quine only if, as I suspect, he wants to *eliminate* the notion of knowledge from the theory of meaning altogether.) But, now, this requirement calls in question the feasibility of any model of understanding, any theory of meaning, according to which the understanding of a sentence consists, in general, of a knowledge of its truth-conditions, when the notion of truth is construed as satisfying the principle of bivalence and as, in general, given independently of our means of recognizing truth. Of the three arguments which I cited as contained in middle and late Wittgenstein, it is the third I wish to stress here. Our language contains many sentences for which we know no procedure, even in principle, which will put us in a position to assert or deny that sentence, at least with full justification. Indeed, for many such sentences, we have no ground for supposing that there necessarily exists any means whereby we could recognize the sentence as true or as false, even means of which we have no effective method of availing ourselves. Hence a notion of truth for such a sentence, taken as subject to the principle of bivalence, cannot be

[4] *Mind and Language*, ed. S. Guttenplan (Oxford: Clarendon Press, 1975), 83–96 (now Essay V in this volume).

equated with the existence of a means of justifying an assertion of it—an equation which sufficed for the distinction between the meaning of a sentence, as a type, and the point of a particular assertion of it. More importantly, a speaker's knowledge of the condition which must, in general, hold for the sentence to be true cannot be taken to consist in his ability to recognize it as true whenever those conditions obtain under which it may be so recognized, and as false when it may be recognized as false, since, by hypothesis, it may be true even in the absence of any such conditions, and he must know the condition for it to be true in those cases also. Therefore, if meaning is use, that is, if the knowledge in which a speaker's understanding of a sentence consists must be capable of being fully manifested by his linguistic practice, it appears that a model of meaning in terms of a knowledge of truth-conditions is possible only if we construe truth in such a way that the principle of bivalence fails; and this means, in effect, some notion of truth under which the truth of a sentence implies the possibility, in principle, of *our* recognizing its truth. It is hard to swallow such a conclusion, because it has profound metaphysical repercussions: it means that we cannot operate, in general, with a picture of our language as bearing a sense that enables us to talk about a determinate, objective reality which renders what we say determinately true or false independently of whether we have the means to recognize its truth or falsity. On the other hand, if the identification of meaning with use does not impose on a theory of meaning the constraints I have suggested, I for one find it difficult to see how it can impose any constraints whatever.

IX

MEANING AND REFERENCE

HILARY PUTNAM

Unclear as it is, the traditional doctrine that the notion "meaning" possesses the extension/intension ambiguity has certain typical consequences. The doctrine that the meaning of a term is a concept carried the implication that meanings are mental entities. Frege, however, rebelled against this "psychologism". Feeling that meanings are *public* property—that the *same* meaning can be "grasped" by more than one person and by persons at different times—he identified concepts (and hence "intensions" or meanings) with abstract entities rather than mental entities. However, "grasping" these abstract entities was still an individual psychological act. None of these philosophers doubted that understanding a word (knowing its intension) was just a matter of being in a certain psychological state (somewhat in the way in which knowing how to factor numbers in one's head is just a matter of being in a certain very complex psychological state).

Secondly, the timeworn example of the two terms 'creature with a kidney' and 'creature with a heart' does show that two terms can have the same extension and yet differ in intension. But it was taken to be obvious that the reverse is impossible: two terms cannot differ in extension and have the same intension. Interestingly, no argument for this impossibility was ever offered. Probably it reflects the tradition of the ancient and medieval philosophers, who assumed that the concept corresponding to a term was just a conjunction of predicates, and hence that the concept corresponding to a term must *always* provide a necessary and sufficient condition for falling into the extension of the term. For philosophers like Carnap, who accepted the verifiability theory of meaning, the concept corresponding to a term provided (in the ideal case, where the term had "complete meaning") a *criterion* for belonging to the extension (not just in the sense of "necessary and sufficient condition", but in the strong sense of *way of recognizing* whether a given thing falls

From *The Journal of Philosophy*, 70/19 (8 November 1973), 699–711. Reprinted by permission of the author and *The Journal of Philosophy*.

into the extension or not). So theory of meaning came to rest on two unchallenged assumptions:

(1) That knowing the meaning of a term is just a matter of being in a certain psychological state (in the sense of "psychological state", in which states of memory and belief are "psychological states"; no one thought that knowing the meaning of a word was a continuous state of consciousness, of course).
(2) That the meaning of a term determines its extension (in the sense that sameness of intension entails sameness of extension).

I shall argue that these two assumptions are not jointly satisfied by *any* notion, let alone any notion of meaning. The traditional concept of meaning is a concept which rests on a false theory.

ARE MEANINGS IN THE HEAD?

For the purpose of the following science-fiction examples, we shall suppose that somewhere there is a planet we shall call Twin Earth. Twin Earth is very much like Earth: in fact, people on Twin Earth even speak *English*. In fact, apart from the differences we shall specify in our science-fiction examples, the reader may suppose that Twin Earth is *exactly* like Earth. He may even suppose that he has a *Doppelgänger*—an identical copy—on Twin Earth, if he wishes, although my stories will not depend on this.

Although some of the people on Twin Earth (say, those who call themselves "Americans" and those who call themselves "Canadians" and those who call themselves "Englishmen", etc.) speak English, there are, not surprisingly, a few tiny differences between the dialects of English spoken on Twin Earth and standard English.

One of the peculiarities of Twin Earth is that the liquid called "water" is not H_2O but a different liquid whose chemical formula is very long and complicated. I shall abbreviate this chemical formula simply as XYZ. I shall suppose that XYZ is indistinguishable from water at normal temperatures and pressures. Also, I shall suppose that the oceans and lakes and seas of Twin Earth contain XYZ and not water, that it rains XYZ on Twin Earth and not water, etc.

If a space ship from Earth ever visits Twin Earth, then the supposition at first will be that 'water' has the same meaning on Earth and on Twin Earth. This supposition will be corrected when it is discovered that "water" on Twin Earth is XYZ, and the Earthian space ship will report somewhat as follows.

"On Twin Earth the word 'water' means XYZ."

Symmetrically, if a space ship from Twin Earth ever visits Earth, then the supposition at first will be that the word 'water' has the same meaning on Twin Earth and on Earth. This supposition will be corrected when it is discovered that "water" on Earth is H_2O, and the Twin Earthian space ship will report:

"On Earth the word 'water' means H_2O."

Note that there is no problem about the extension of the term 'water': the word simply has two different meanings (as we say); in the sense in which it is used on Twin Earth, the sense of $water_{TE}$, what *we* call "water" simply isn't water, while in the sense in which it is used on Earth, the sense of $water_E$, what the Twin Earthians call "water" simply isn't water. The extension of 'water' in the sense of $water_E$ is the set of all wholes consisting of H_2O molecules, or something like that; the extension of water in the sense of $water_{TE}$ is the set of all wholes consisting of XYZ molecules, or something like that.

Now let us roll the time back to about 1750. The typical Earthian speaker of English did not know that water consisted of hydrogen and oxygen, and the typical Twin Earthian speaker of English did not know that "water" consisted of XYZ. Let $Oscar_1$ be such a typical Earthian English speaker, and let $Oscar_2$ be his counterpart on Twin Earth. You may suppose that there is no belief that $Oscar_1$ had about water that $Oscar_2$ did not have about "water". If you like, you may even suppose that $Oscar_1$ and $Oscar_2$ were exact duplicates in appearance, feelings, thoughts, interior monologue, etc. Yet the extension of the term 'water' was just as much H_2O on Earth in 1750 as in 1950; and the extension of the term 'water' was just as much XYZ on Twin Earth in 1750 as in 1950. $Oscar_1$ and $Oscar_2$ understood the term 'water' differently in 1750 *although they were in the same psychological state*, and although, given the state of science at the time, it would have taken their scientific communities about fifty years to discover that they understood the term 'water' differently. Thus the extension of the term 'water' (and, in fact, its "meaning" in the intuitive preanalytical usage of that term) is *not* a function of the psychological state of the speaker by itself.[1]

But, it might be objected, why should we accept it that the term 'water' had the same extension in 1750 and in 1950 (on both Earths)? Suppose I point to a glass of water and say "this liquid is called water." My "ostensive definition" of water has the following empirical pre-

[1] See n. 2 below, and the corresponding text.

supposition: that the body of liquid I am pointing to bears a certain sameness relation (say, *x is the same liquid as y*, or *x is the same*$_L$ *as y*) to most of the stuff I and other speakers in my linguistic community have on other occasions called "water". If this presupposition is false because, say, I am—unknown to me—pointing to a glass of gin and not a glass of water, then I do not intend my ostensive definition to be accepted. Thus the ostensive definition conveys what might be called a "defeasible" necessary and sufficient condition: the necessary and sufficient condition for being water is bearing the relation *same*$_L$ to the stuff in the glass; but this is the necessary and sufficient condition only if the empirical presupposition is satisfied. If it is not satisfied, then one of a series of, so to speak, "fallback" conditions becomes activated.

The key point is that the relation *same*$_L$ is a *theoretical* relation: whether something is or is not the same liquid as *this* may take an indeterminate amount of scientific investigation to determine. Thus, the fact that an English speaker in 1750 might have called XYZ "water", whereas he or his successors would not have called XYZ water in 1800 or 1850 does not mean that the "meaning" of 'water' changed for the average speaker in the interval. In 1750 or in 1850 or in 1950 one might have pointed to, say, the liquid in Lake Michigan as an example of "water". What changed was that in 1750 we would have mistakenly thought that XYZ bore the relation *same*$_L$ to the liquid in Lake Michigan, whereas in 1800 or 1850 we would have known that it did not.

Let us now modify our science-fiction story. I shall suppose that molybdenum pots and pans *can't* be distinguished from aluminium pots and pans save by an expert. (This could be true for all I know, and, a fortiori, it could be true for all I know by virtue of "knowing the meaning" of the words *aluminium* and *molybdenum*.) We will now suppose that molybdenum is as common on Twin Earth as aluminium is on Earth, and that aluminium is as rare on Twin Earth as molybdenum is on Earth. In particular, we shall assume that "aluminium" pots and pans are made of molybdenum on Twin Earth. Finally, we shall assume that the words 'aluminium' and 'molybdenum' are *switched* on Twin Earth: 'aluminium' is the name of *molybdenum*, and 'molybdenum' is the name of *aluminium*. If a space ship from Earth visited Twin Earth, the visitors from Earth probably would not suspect that the "aluminium" pots and pans on Twin Earth were not made of aluminium, especially when the Twin Earthians *said* they were. But there is one important difference between the two cases. An Earthian metallurgist could tell very easily that "aluminium" was molybdenum, and a Twin Earthian metallurgist could tell equally easily that aluminium was "molybdenum". (The shudder quotes in the preceding sentence indicate Twin Earthian usages.) Whereas in 1750 no one on either Earth or Twin Earth could have distinguished water from "water", the confusion of aluminium with "aluminium" involves only a part of the linguistic communities involved.

This example makes the same point as the preceding example. If Oscar$_1$ and Oscar$_2$ are standard speakers of Earthian English and Twin Earthian English, respectively, and neither is chemically or metallurgically sophisticated, then there may be no difference at all in their psychological states when they use the word 'aluminium'; nevertheless, we have to say that 'aluminium' has the extension *aluminium* in the idiolect of Oscar$_1$ and the extension *molybdenum* in the idiolect of Oscar$_2$. (Also we have to say that Oscar$_1$ and Oscar$_2$ mean different things by 'aluminium'; that 'aluminium' has a different meaning on Earth than it does on Twin Earth, etc.) Again we see that the psychological state of the speaker does *not* determine the extension (*or* the "meaning", speaking preanalytically) of the word.

Before discussing this example further, let me introduce a *non*-science-fiction example. Suppose you are like me and cannot tell an elm from a beech tree. We still say that the extension of 'elm' in my idiolect is the same as the extension of 'elm' in anyone else's, viz., the set of all elm trees, and that the set of all beech trees is the extension of 'beech' in *both* of our idiolects. Thus 'elm' in my idiolect has a different extension from 'beech' in your idiolect (as it should). Is it really credible that this difference in extension is brought about by some difference in our *concepts*? My *concept* of an elm tree is exactly the same as my concept of a beech tree (I blush to confess). If someone heroically attempts to maintain that the difference between the extension of 'elm' and the extension of 'beech' in *my* idiolect is explained by a difference in my psychological state, then we can always refute him by constructing a "Twin Earth" example—just let the words 'elm' and 'beech' be switched on Twin Earth (the way 'aluminium' and "molybdenum" were in the previous example). Moreover, suppose I have a *Doppelgänger* on Twin Earth who is molecule for molecule "identical" with me. If you are a dualist, then also suppose my *Doppelgänger* thinks the same verbalized thoughts I do, has the same sense data, the same dispositions, etc. It is absurd to think *his* psychological state is one bit different from mine: yet he "means" *beech* when he says "elm", and *I* "mean" *elm* when I say "elm". Cut the pie any way you like, "meanings" just ain't in the *head*!

A SOCIOLINGUISTIC HYPOTHESIS

The last two examples depend upon a fact about language that seems, surprisingly, never to have been pointed out: that there is *division of linguistic labour*. We could hardly use such words as 'elm' and 'aluminium' if no one possessed a way of recognizing elm trees and aluminium metal; but not everyone to whom the distinction is important has to be able to make the distinction. Let us shift the example; consider *gold*. Gold is important for many reasons: it is a precious metal; it is a monetary metal; it has symbolic value (it is important to most people that the "gold" wedding ring they wear *really* consist of gold and not just *look* gold); etc. Consider our community as a "factory": in this "factory" some people have the "job" of *wearing gold wedding rings*; other people have the "job" of selling gold wedding rings; still other people have the job of *telling whether or not something is really gold*. It is not at all necessary or efficient that everyone who wears a gold ring (or a gold cufflink, etc.), or discusses the "gold standard", etc., engage in buying and selling gold. Nor is it necessary or efficient that everyone who buys and sells gold be able to tell whether or not something is really gold in a society where this form of dishonesty is uncommon (selling fake gold) and in which one can easily consult an expert in case of doubt. And it is *certainly* not necessary or efficient that everyone who has occasion to buy or wear gold be able to tell with any reliability whether or not something is really gold.

The foregoing facts are just examples of mundane division of labour (in a wide sense). But they engender a division of linguistic labour: everyone to whom gold is important for any reason has to *acquire* the word 'gold'; but he does not have to acquire the *method of recognizing* whether something is or is not gold. He can rely on a special subclass of speakers. The features that are generally thought to be present in connection with a general name—necessary and sufficient conditions for membership in the extension, ways of recognizing whether something is in the extension, etc.—are all present in the linguistic community *considered as a collective body*; but that collective body divides the "labour" of knowing and employing these various parts of the "meaning" of 'gold'.

This division of linguistic labour rests upon and presupposes the division of *non*-linguistic labour, of course. If only the people who know how to tell whether some metal is really gold or not have any reason to have the word 'gold' in their vocabulary, then the word 'gold' will be as the word 'water' was in 1750 with respect to that subclass of speakers, and the other speakers just won't acquire it at all. And some words do not exhibit any division of linguistic labour: 'chair', for example. But with the increase of division of labour in the society and the rise of science, more and more words begin to exhibit this kind of division of labour. 'Water', for example, did not exhibit it at all before the rise of chemistry. Today it is obviously necessary for every speaker to be able to recognize water (reliably under normal conditions), and probably most adult speakers even know the necessary and sufficient condition "water is H_2O", but only a few adult speakers could distinguish water from liquids that superficially resembled water. In case of doubt, other speakers would rely on the judgement of these "expert" speakers. Thus the way of recognizing possessed by these "expert" speakers is also, through them, possessed by the collective linguistic body, even though it is not possessed by each individual member of the body, and in this way the most *recherché* fact about water may become part of the *social* meaning of the word although unknown to almost all speakers who acquire the word.

It seems to me that this phenomenon of division of linguistic labour is one that it will be very important for sociolinguistics to investigate. In connection with it, I should like to propose the following hypothesis:

> HYPOTHESIS OF THE UNIVERSALITY OF THE DIVISION OF LINGUISTIC LABOUR: Every linguistic community exemplifies the sort of division of linguistic labour just described; that is, it possesses at least some terms whose associated "criteria" are known only to a subset of the speakers who acquire the terms, and whose use by the other speakers depends upon a structured co-operation between them and the speakers in the relevant subsets.

It is easy to see how this phenomenon accounts for some of the examples given above of the failure of the assumptions (1 and 2). When a term is subject to the division of linguistic labour, the "average" speaker who acquires it does not acquire anything that fixes its extension. In particular, his individual psychological state *certainly* does not fix its extension; it is only the sociolinguistic state of the collective linguistic body to which the speaker belongs that fixes the extension.

We may summarize this discussion by pointing out that there are two sorts of tools in the world: there are tools like a hammer or a screwdriver which can be used by one person; and there are tools like a steamship which require the co-operative activity of a number of persons to use.

Words have been thought of too much on the model of the first sort of tool.

INDEXICALITY AND RIGIDITY

The first of our science-fiction examples—'water' on Earth and on Twin Earth in 1750—does not involve division of linguistic labour, or at least does not involve it in the same way the examples of 'aluminium' and 'elm' do. There were not (in our story, anyway) any "experts" on water on Earth in 1750, nor any experts on "water" on Twin Earth. The example *does* involve things which are of fundamental importance to the theory of reference and also to the theory of necessary truth, which we shall now discuss.

Let W_1 and W_2 be two possible worlds in which I exist and in which this glass exists and in which I am giving a meaning explanation by pointing to this glass and saying "This is water." Let us suppose that in W_1 the glass is full of H_2O and in W_2 the glass is full of XYZ. We shall also suppose that W_1 is the *actual* world, and that XYZ is the stuff typically called "water" in the world W_2 (so that the relation between English speakers in W_1 and English speakers in W_2 is exactly the same as the relation between English speakers on Earth and English speakers on Twin Earth). Then there are two theories one might have concerning the meaning of 'water':

(1) One might hold that 'water' was *world-relative* but *constant* in meaning (i.e., the word has a constant relative meaning). On this theory, 'water' means the same in W_1 and W_2; it's just that water is H_2O in W_1, and water is XYZ in W_2.
(2) One might hold that water is H_2O in all worlds (the stuff called "water" in W_2 isn't water), but 'water' doesn't have the same meaning in W_1 and W_2.

If what was said before about the Twin Earth case was correct, then (2) is clearly the correct theory. When I say "*this* (liquid) is water", the "this" is, so to speak, a *de re* "this"—i.e., the force of my explanation is that "water" is whatever bears a certain equivalence relation (the relation we called "$same_L$" above) to the piece of liquid referred to as "this" *in the actual world*.

We might symbolize the difference between the two theories as a "scope" difference in the following way. On theory (1), the following is true:

(1′) (For every world W) (For every x in W) (x is water $\equiv x$ bears $same_L$ to the entity referred to as "this" in W)

while on theory (2):

(2′) (For every world W) (For every x in W) (x is water $\equiv x$ bears $same_L$ to the entity referred to as "this" *in the actual world* W_1)

I call this a "scope" difference because in (1′) 'the entity referred to as "this"' is within the scope of 'For every world W'—as the qualifying phrase 'in W' makes explicit—whereas in (2′) 'the entity referred to as "this"' means "the entity referred to as 'this' *in the actual world*", and has thus a reference *independent* of the bound variable 'W'.

Kripke calls a designator "rigid" (in a given sentence) if (in that sentence) it refers to the same individual in every possible world in which the designator designates. If we extend this notion of rigidity to substance names, then we may express Kripke's theory and mine by saying that the term 'water' is *rigid*.

The rigidity of the term 'water' follows from the fact that when I give the "ostensive definition": "*this* (liquid) is water", I intend (2′) and not (1′).

We may also say, following Kripke, that when I give the ostensive definition "*this* (liquid) is water", the demonstrative 'this' is *rigid*.

What Kripke was the first to observe is that this theory of the meaning (or "use", or whatever) of the word 'water' (and other natural kind terms as well) has startling consequences for the theory of necessary truth.

To explain this, let me introduce the notion of a *cross-world relation*. A two-term relation R will be called *cross-world* when it is understood in such a way that its extension is a set of ordered pairs of individuals *not all in the same possible world*. For example, it is easy to understand the relation *same height as* as a cross-world relation: just understand it so that, e.g., if x is an individual in a world W_1 who is 5 feet tall (in W_1) and y is an individual in W_2 who is 5 feet tall (in W_2), then the ordered pair x,y belongs to the extension of *same height as*. (Since an individual may have different heights in different possible worlds in which that same individual exists, strictly speaking, it is not the ordered pair x,y that constitutes an element of the extension of *same height as*, but rather the ordered pair x-*in-world*-W_1, y-*in-world*-W_2.)

Similarly, we can understand the relation $same_L$ (same liquid as) as a cross-world relation by understanding it so that a liquid in world W_1 which has the same important physical properties (in W_1) that a liquid in W_2 possesses (in W_2) bears $same_L$ to the latter liquid.

Then the theory we have been presenting may be summarized by saying that an entity *x*, in an arbitrary possible world, is *water* if and only if it bears the relation $same_L$ (construed as a cross-world relation) to the stuff *we* call "water" in the actual world.

Suppose, now, that I have not yet discovered what the important physical properties of water are (in the actual world)—i.e., I don't yet know that water is H_2O. I may have ways of *recognizing* water that are successful (of course, I may make a small number of mistakes that I won't be able to detect until a later stage in our scientific development), but not know the microstructure of water. If I agree that a liquid with the superficial properties of "water" but a different microstructure *isn't really water*, then my ways of recognizing water cannot be regarded as an analytical specification of what *it is to be* water. Rather, the operational definition, like the ostensive one, is simply a way of pointing out a standard—pointing out the stuff *in the actual world* such that, for *x* to be water, in *any* world, is for *x* to bear the relation $same_L$ to the *normal* members of the class of *local* entities that satisfy the operational definition. "Water" on Twin Earth is not water, even if it satisfies the operational definition, because it doesn't bear $same_L$ to the *local* stuff that satisfies the operational definition, and local stuff that satisfies the operational definition but has a microstructure different from the rest of the local stuff that satisfies the operational definition isn't water either, because it doesn't bear $same_L$ to the *normal* examples of the local "water".

Suppose, now, that I discover the microstructure of water—that water is H_2O. At this point I will be able to say that the stuff on Twin Earth that I earlier *mistook* for water isn't really water. In the same way, if you describe, not another planet in the actual universe, but another possible universe in which there is stuff with the chemical formula XYZ which passes the "operational test" for *water*, we shall have to say that that stuff isn't water but merely XYZ. You will not have described a possible world in which "water is XYZ", but merely a possible world in which there are lakes of XYZ, people drink XYZ (and not water), or whatever. In fact, once we have discovered the nature of water, nothing counts as a possible world in which water doesn't have that nature. Once we have discovered that water (in the actual world) is H_2O, *nothing counts as a possible world in which water isn't* H_2O.

On the other hand, we can perfectly well imagine having experiences that would convince us (and that would make it rational to believe that) water *isn't* H_2O. In that sense, it is conceivable that water isn't H_2O. It is conceivable but it isn't possible! Conceivability is no proof of possibility.

Kripke refers to statements that are rationally unrevisable (assuming there are such) as *epistemically necessary*. Statements that are true in all possible worlds he refers to simply as necessary (or sometimes as "metaphysically necessary"). In this terminology, the point just made can be restated as: a statement can be (metaphysically) necessary and epistemically contingent. Human intuition has no privileged access to metaphysical necessity.

In this essay, our interest is in theory of meaning, however, and not in theory of necessary truth. Words like 'now', 'this', 'here' have long been recognized to be *indexical*, or *token-reflexive*—i.e., to have an extension which varies from context to context or token to token. For these words, no one has ever suggested the traditional theory that "intension determines extension". To take our Twin Earth example: if I have a *Doppelgänger* on Twin Earth, then when I think "I have a headache," *he* thinks "I have a headache." But the extension of the particular token of 'I' in his verbalized thought is himself (or his unit class, to be precise), while the extension of the token of 'I' in *my* verbalized thought is *me* (or my unit class, to be precise). So the same word, 'I', has two different extensions in two different idiolects; but it does not follow that the concept I have of myself is in any way different from the concept my *Doppelgänger* has of himself.

Now then, we have maintained that indexicality extends beyond the *obviously* indexical words and morphemes (e.g., the tenses of verbs). Our theory can be summarized as saying that words like 'water' have an unnoticed indexical component: "water" is stuff that bears a certain similarity relation to the water *around here*. Water at another time or in another place or even in another possible world has to bear the relation same_{L} to *our* "water" *in order to be water*. Thus the theory that (1) words have "intensions", which are something like concepts associated with the words by speakers; and (2) intension determines extension—cannot be true of natural kind words like 'water' for the same reason it cannot be true of obviously indexical words like 'I'.

The theory that natural kind words like 'water' are indexical leaves it open, however, whether to say that 'water' in the Twin Earth dialect of English has the same *meaning* as 'water' in the Earth dialect and a different extension—which is what we normally say about 'I' in different idiolects—thereby giving up the doctrine that "meaning (intension) determines extension", or to say, as we have chosen to do, that difference in extension is *ipso facto* a difference in meaning for natural kind words, thereby giving up the doctrine that meanings are concepts, or, indeed, mental entities of *any* kind.[2]

It should be clear, however, that Kripke's doctrine that natural kind words are rigid designators and our doctrine that they are indexical are but two ways of making the same point.

We have now seen that the extension of a term is not fixed by a concept that the individual speaker has in his head, and this is true both because extension is, in general, determined *socially*—there is division of linguistic labour as much as of "real" labour—and because extension is, in part, determined *indexically*. The extension of our terms depends upon the actual nature of the particular things that serve as paradigms, and this actual nature is not, in general, fully known to the speaker. Traditional semantic theory leaves out two contributions to the determination of reference—the contribution of society and the contribution of the real world; a better semantic theory must encompass both.

[2] Our reasons for rejecting the first option—to say that 'water' has the same meaning on Earth and on Twin Earth, while giving up the doctrine that meaning determines references—are presented in "The Meaning of 'Meaning'". They may be illustrated thus: Suppose 'water' has the same meaning on Earth and on Twin Earth. Now, let the word 'water' become phonemically different on Twin Earth—say, it becomes 'quaxel'. Presumably, this is not a change in meaning *per se*, on any view. So 'water' and 'quaxel' have the same meaning (although they refer to different liquids). But this is highly counterintuitive. Why not say, then, that 'elm' in my idiolect has the same meaning as 'beech' in your idiolect, although they refer to different trees?

X

IDENTITY AND NECESSITY

SAUL KRIPKE

A problem which has arisen frequently in contemporary philosophy is: 'How are *contingent* identity statements possible?" This question is phrased by analogy with the way Kant phrased his question "How are synthetic a priori judgements possible?" In both cases, it has usually been taken for granted in the one case by Kant that synthetic a priori judgements were possible, and in the other case in contemporary philosophical literature that contingent statements of identity are possible. I do not intend to deal with the Kantian question except to mention this analogy: after a rather thick book was written trying to answer the question how synthetic a priori judgements were possible, others came along later who claimed that the solution to the problem was that synthetic a priori judgements were, of course, impossible and that a book trying to show otherwise was written in vain. I will not discuss who was right on the possibility of synthetic a priori judgements. But in the case of contingent statements of identity, most philosophers have felt that the notion of a contingent identity statement ran into something like the following paradox. An argument like the following can be given against the possibility of contingent identity statements:[1] First, the law of the substitutivity of identity says that, for any objects x and y, if x is identical to y, then if x has a certain property F, so does y:

From *Identity and Individuation*, ed. Milton K. Munitz (New York University Press, 1971), 135–64. Reprinted by permission of the author.

[1] This paper was presented orally, without a written text, to the New York University lecture series on identity which makes up the volume *Identity and Individuation*. The lecture was taped, and the present essay represents a transcription of these tapes, edited only slightly with no attempt to change the style of the original. If the reader imagines the sentences of this essay as being delivered, extemporaneously, with proper pauses and emphases, this may facilitate his comprehension. Nevertheless, there may still be passages which are hard to follow, and the time allotted necessitated a condensed presentation of the argument. (A longer version of some of these views, still rather compressed and still representing a transcript of oral remarks, has appeared in *Semantics of Natural Language*, ed. By Donald Davidson and Gilbert Harman (Dordrecht: D. Reidel, 1972).) Occasionally, reservations, amplifications and gratifications of my remarks had to be repressed, especially in the discussion of theoretical identification and the mind–body problem. The footnotes, which were added to the original, would have become even more unwieldy if this had not been done.

(1) $(x)(y)[(x=y) \supset (Fx \supset Fy)]$

On the other hand, every object surely is necessarily self-identical:

(2) $(x)\Box(x=x)$

But

(3) $(x)(y)(x=y) \supset [\Box(x=x) \supset \Box(x=y)]$

is a substitution instance of (1), the substitutivity law. From (2) and (3), we can conclude that, for every x and y, if x equals y, then, it is necessary that x equals y:

(4) $(x)(y)((x=y) \supset \Box(x=y))$

This is because the clause $\Box(x=x)$ of the conditional drops out because it is known to be true.

This is an argument which has been stated many times in recent philosophy. Its conclusion, however, has often been regarded as highly paradoxical. For example, David Wiggins, in his paper, 'Identity-Statements', says,

> Now there undoubtedly exist contingent identity-statements. Let $a=b$ be one of them. From its simple truth and (5) [=(4) above] we can derive '$\Box(a=b)$'. But how then can there be any contingent identity-statements?[2]

He then says that five various reactions to this argument are possible, and rejects all of these reactions, and reacts himself. I do not want to discuss all the possible reactions to this statement, except to mention the second of those Wiggins rejects. This says,

> We might accept the result and plead that provided '*a*' and '*b*' are proper names nothing is amiss. The consequence of this is that no contingent identity-statements can be made by means of proper names.

And then he says that he is discontented with this solution and many other philosophers have been discontented with this solution, too, while still others have advocated it.

What makes the statement (4) seem surprising? It says, for any objects x and y, if x is y, then it is necessary that x is y. I have already mentioned that someone might object to this argument on the grounds that premise

[2] R. J. Butler (ed.), *Analytical Philosophy*, Second Series (Oxford: Blackwell, 1965), 41. [Ed. note: David Wiggins recanted the defence of contingent identity in *Identity and Spatio-Temporal Continuity* (Oxford: Blackwell, 1967): see n. 7 there. He defended the necessity of identity in 'On Sentence Sense, Word Sense and Difference of Word Sense', in D. Steinberg and L. Jakobovits (eds.), *Semantics: An Interdisciplinary Reader* (Cambridge: Cambridge University Press, 1971).]

(2) is already false, that it is not the case that everything is necessarily self-identical. Well, for example, am I myself necessarily self-identical? Someone might argue that in some situations which we can imagine I would not even have existed and therefore the statement "Saul Kripke is Saul Kripke" would have been false or it would not be the case that I was self-identical. Perhaps, it would have been neither true nor false, in such a world, to say that Saul Kripke is self-identical. Well, that may be so, but really it depends on one's philosophical view of a topic that I will not discuss, that is, what is to be said about truth values of statements mentioning objects that do not exist in the actual world or any given possible world or counterfactual situation. Let us interpret necessity here weakly. We can count statements as necessary if whenever the objects mentioned therein exist, the statement would be true. If we wished to be very careful about this, we would have to go into the question of existence as a predicate and ask if the statement can be reformulated in the form: For every x it is necessary that, if x exists, then x is self-identical. I will not go into this particular form of subtlety here because it is not going to be relevant to my main theme. Nor am I really going to consider formula (4). Anyone who believes formula (2) is, in my opinion, committed to formula (4). If x and y are the same things and we can talk about modal properties of an object at all, that is, in the usual parlance, we can speak of modality *de re* and an object *necessarily* having certain properties as such, then formula (1), I think, has to hold. Where x is any property at all, including a property involving modal operators, and if x and y are the same object and x had a certain property F, then y has to have the same property F. And this is so even if the property F is itself of the form of necessarily having some other property G, in particular that of necessarily being identical to a certain object. Well, I will not discuss the formula (4) itself because by itself it does not assert, of any particular true statement of identity, that it is necessary. It does not say anything about *statements* at all. It says for every *object* x and *object* y, if x and y are the same object, then it is necessary that x and y are the same object. And this, I think, if we think about it (anyway, if someone does not think so, I will not argue for it here), really amounts to something very little different from the statement (2). Since x, by definition of identity, is the only object identical with x, "$(y)(y = x \supset Fy)$" seems to me to be little more than a garrulous way of saying 'Fx', and thus $(x)(y)(y = x \supset Fx)$ says the same as $(x)Fx$ no matter what 'F' is—in particular, even if 'F' stands for the property of necessary identity with x. So if x has this property (of necessary identity with x), trivially everything identical with x has it, as (4) asserts.

But, from statement (4) one may apparently be able to deduce that various particular statements of identity must be necessary and this is then supposed to be a very paradoxical consequence.

Wiggins says, "Now there undoubtedly exist contingent identity-statements." One example of a contingent identity statement is the statement that the first Postmaster General of the United States is identical with the inventor of bifocals, or that both of these are identical with the man claimed by the *Saturday Evening Post* as its founder (*falsely* claimed, I gather, by the way). Now some such statements are plainly contingent. It plainly is a contingent fact that one and the same man both invented bifocals and took on the job of Postmaster General of the United States. How can we reconcile this with the truth of statement (4)? Well, that, too, is an issue I do not want to go into in detail except to be very dogmatic about it. It was I think settled quite well by Bertrand Russell in his notion of the scope of a description. According to Russell, one can, for example, say with propriety that the author of *Hamlet* might not have written *Hamlet*, or even that the author of *Hamlet* might not have been the author of *Hamlet*. Now here, of course, we do not deny the necessity of the identity of an object with itself; but we say it is true concerning a certain man that he in fact was the unique person to have written *Hamlet* and secondly that the man, who in fact was the man who wrote *Hamlet*, might not have written *Hamlet*. In other words, if Shakespeare had decided not to write tragedies, he might not have written *Hamlet*. Under these circumstances, the man who in fact wrote *Hamlet* would not have written *Hamlet*. Russell brings this out by saying that in such a statement, the first occurrence of the description "the author of *Hamlet*" has large scope.[3] That is, we say "The author of *Hamlet* has the following property: that he might not have written *Hamlet*." We *do not* assert that the following statement might have been the case, namely that the author of *Hamlet* did not write *Hamlet*, for that is not true. That would be to say that it might have been the case that someone wrote *Hamlet* and yet did not write *Hamlet*, which would be a contradiction. Now, aside from the details of Russell's particular formulation of it, which depends on his theory of descriptions, this seems to be the distinction that any theory of descriptions has to make. For example, if someone were to meet the President of Harvard and take him to be a Teaching Fellow, he might say: "I took the President of Harvard for a Teaching Fellow." By this he does not mean that he took the proposition "The President of Harvard is

[3] The second occurrence of the description has small scope.

a Teaching Fellow" to be true. He could have meant this, for example, had he believed that some sort of democratic system had gone so far at Harvard that the President of it decided to take on the task of being a Teaching Fellow. But that probably is not what he means. What he means instead, as Russell points out, is "Someone is President of Harvard and I took him to be a Teaching Fellow." In one of Russell's examples someone says, "I thought your yacht is much larger than it is." And the other man replies, "No, my yacht is not much larger than it is."

Provided that the notion of modality *de re*, and thus of quantifying into modal contexts, makes any sense at all, we have quite an adequate solution to the problem of avoiding paradoxes if we substitute descriptions for the universal quantifiers in (4) because the only consequence we will draw,[4] for example, in the bifocals case, is that there is a man who both happened to have invented bifocals and happened to have been the first Postmaster General of the United States, and is necessarily self-identical. There is an object x such that x invented bifocals, and as a matter of contingent fact an object y, such that y is the first Postmaster General of the United States, and finally, it is necessary, that x is y. What are x and y here? Here, x and y are both Benjamin Franklin, and it can certainly be necessary that Benjamin Franklin is identical with himself. So, there is no problem in the case of descriptions if we accept Russell's notion of scope.[5] And I just dogmatically want to drop that

[4] In Russell's theory, $F(\iota xGx)$ follows from $(x)Fx$ and $(\exists!x)$ Gx, provided that the description in $F(\iota xGx)$ has the entire context for its scope (in Russell's 1905 terminology, has a 'primary occurrence'). Only then is $F(\iota xGx)$ 'about' the denotation of 'ιxGx'. Applying this rule to (4), we get the results indicated in the text. Notice that, in the ambiguous form $\Box(\iota xGx = \iota xHx)$, if one or both of the descriptions have 'primary occurrences' the formula does not assert the necessity of $\iota xGx = \iota xHx$; if both have secondary occurrences, it does. Thus in a language without explicit scope indicators, descriptions must be construed with the smallest possible scope—only then will $\sim A$ be the negation of A, $\Box A$ the necessitation of A, and the like.

[5] An earlier distinction with the same purpose was, of course, the medieval one of *de dicto–de re*. That Russell's distinction of scope eliminates modal paradoxes has been pointed out by many logicians, especially Smullyan.

So as to avoid misunderstanding, let me emphasize that I am of course not asserting that Russell's notion of scope solves Quine's problem of 'essentialism'; what it does show, especially in conjunction with modern model-theoretic approaches to modal logic, is that quantified modal logic need not deny the truth of all instances of $(x)(y)(x = y \cdot \supset \cdot Fx \supset Fy)$, nor of all instances of '$(x)(Gx \supset Ga)$' (where 'a' is to be replaced by a nonvacuous definite description whose scope is all of 'Ga'), in order to avoid making it a necessary truth that one and the same man invented bifocals and headed the original Postal Department. Russell's contextual definition of description need not be adopted in order to ensure these results; but other logical theories, Fregean or other, which take descriptions as primitive must somehow express the same logical facts. Frege showed that a simple, non-iterated context containing a definite description with small scope, which cannot be interpreted as being 'about' the denotation of the description, can be interpreted as about its 'sense'. Some logicians have

question here and go on to the question about names which Wiggins raises. And Wiggins says he might accept the result and plead that, provided *a* and *b* are proper names, nothing is amiss. And then he rejects this.

Now what is the special problem about proper names? At least if one is not familiar with the philosophical literature about this matter, one naïvely feels something like the following about proper names. First, if someone says "Cicero was an orator", then he uses the name 'Cicero' in that statement simply to pick out a certain object and then to ascribe a certain property to the object, namely, in this case, he ascribes to a certain man the property of having been an orator. If someone else uses another name, such as, say, 'Tully', he is still speaking about the same man. One ascribes the same property, if one says "Tully is an orator", to the same man. So to speak, the fact, or state of affairs, represented by the statement is the same whether one says "Cicero is an orator" or one says "Tully is an orator." It would, therefore, seem that the function of names is *simply* to refer, and not to describe the objects so named by such properties as "being the inventor of bifocals" or "being the first Postmaster General". It would seem that Leibniz' law and the law (1) should not only hold in the universally quantified form, but also in the form "if $a = b$ and Fa, then Fb", wherever 'a' and 'b' stand in place of names and 'F' stands in place of a predicate expressing a genuine property of the object:

$$(a = b \cdot Fa) \supset Fb$$

We can run the same argument through again to obtain the conclusion where 'a' and 'b' replace any names, "If $a = b$, then necessarily $a = b$." And so, we could venture this conclusion: that whenever 'a' and 'b' are proper names, if *a* is *b*, that it is necessary that *a* is *b*. Identity statements between proper names have to be necessary if they are going to be true at all. This view in fact has been advocated, for example, by Ruth Barcan Marcus in a paper of hers on the philosophical interpretation of modal logic.[6] According to this view, whenever, for example, someone

been interested in the question of the conditions under which, in an intensional context, a description with small scope is equivalent to the same one with large scope. One of the virtues of a Russellian treatment of descriptions in modal logic is that the answer (roughly that the description be a 'rigid designator' in the sense of this lecture) then often follows from the other postulates for quantified modal logic: no special postulates are needed, as in Hintikka's treatment. Even if descriptions are taken as primitive, special postulation of when scope is irrelevant can often be deduced from more basic axioms.

[6] 'Modalities and Intensional Languages', *Boston Studies in the Philosophy of Science*, i (New York: Humanities Press, 1963), 71 ff. See also the 'Comments' by Quine and the ensuing discussion.

makes a correct statement of identity between two names, such as, for example, that Cicero is Tully, his statement has to be necessary if it is true. But such a conclusion *seems* plainly to be false. (I, like other philosophers, have a habit of understatement in which "it seems plainly false" means "it is plainly false." Actually, I think the view is true, though not quite in the form defended by Mrs Marcus.) At any rate, it seems plainly false. One example was given by Professor Quine in his reply to Professor Marcus at the symposium: "I think I see trouble anyway in the contrast between proper names and descriptions as Professor Marcus draws it. The paradigm of the assigning of proper names is tagging. We may tag the planet Venus some fine evening with the proper name 'Hesperus'. We may tag the same planet again someday before sun rise with the proper name 'Phosphorus'." (Quine thinks that something like that actually was done once.) "When, at last, we discover that we have tagged the same planet twice, our discovery is empirical, and not because the proper names were descriptions." According to what we are told, the planet Venus seen in the morning was originally thought to be a star and was called "the Morning Star", or (to get rid of any question of using a description) was called 'Phosphorus'. One and the same planet, when seen in the evening, was thought to be another star, the Evening Star, and was called "Hesperus". Later on, astronomers discovered that Phosphorus and Hesperus were one and the same. Surely no amount of a priori ratiocination on their part could conceivably have made it possible for them to deduce that Phosphorus is Hesperus. In fact, given the information they had, it might have turned out the other way. Therefore, it is argued, the statement 'Hesperus is Phosphorus' has to be an ordinary contingent, empirical truth, one which might have come out otherwise, and so the view that true identity statements between names are necessary has to be false. Another example which Quine gives in *Word and Object* is taken from Professor Schrödinger, the famous pioneer of quantum mechanics: a certain mountain can be seen from both Tibet and Nepal. When seen from one direction it was called 'Gaurisanker'; when seen from another direction, it was called 'Everest'; and then, later on, the empirical discovery was made that Gaurisanker *is* Everest. (Quine further says that he gathers the example is actually geographically incorrect. I guess one should not rely on physicists for geographical information.)

Of course, one possible reaction to this argument is to deny that names like 'Cicero', 'Tully', 'Gaurisanker', and 'Everest' really are proper names. Look, someone might say (someone has said it: his name was 'Bertrand Russell'), just because statements like "Hesperus is

Phosphorus" and "Gaurisanker is Everest" are contingent, we can see that the names in question are not really purely referential. You are not, in Mrs Marcus's phrase, just 'tagging' an object; you are actually describing it. What does the contingent fact that Hesperus is Phosphorus amount to? Well, it amounts to the fact that *the* star in a certain portion of the sky in the evening is *the* star in a certain portion of the sky in the morning. Similarly, the contingent fact that Gaurisanker is Everest amounts to the fact that the mountain viewed from such and such an angle in Nepal is the mountain viewed from such and such another angle in Tibet. Therefore, such names as 'Hesperus' and 'Phosphorus' can only be abbreviations for descriptions. The term 'Phosphorus' *has* to mean "the star seen . . .", or (let us be cautious because it actually turned out not to be a star), "the *heavenly body* seen from such and such a position at such and such a time in the morning", and the name 'Hesperus' has to mean "the heavenly body seen in such and such a position at such and such a time in the evening." So, Russell concludes, if we want to reserve the term "name" for things which really just name an object without describing it, the only real proper names we can have are names of our own immediate sense data, objects of our own 'immediate acquaintance'. The only such names which occur in language are demonstratives like "this" and "that". And it is easy to see that this requirement of necessity of identity, understood as exempting identities between names from all imaginable doubt, can indeed be guaranteed only for demonstrative names of immediate sense data; for only in such cases can an identity statement between two different names have a general immunity from Cartesian doubt. There are some other things Russell has sometimes allowed as objects of acquaintance, such as one's self; we need not go into details here. Other philosophers (for example, Mrs Marcus in her reply, at least in the verbal discussion as I remember it—I do not know if this got into print, so perhaps this should not be 'tagged' on her[7]) have said, "If names are really just tags, genuine tags, then a good dictionary should be able to tell us that they are names of the same object." You have an object *a* and an object *b* with names 'John' and 'Joe'. Then, according to Mrs Marcus, a dictionary should be able to tell you whether or not 'John' and 'Joe' are names of the same object. Of course, I do not know what ideal dictionaries should do, but ordinary proper names do not seem to satisfy this requirement. You certainly *can*, in the case of ordinary proper names, make quite empirical discoveries that, let's say,

[7] It should. See her remark on p. 115, *Boston Studies in the Philosophy of Science*, i, in the discussion following the papers.

Hesperus is Phosphorus, though we thought otherwise. We can be in doubt as to whether Gaurisanker is Everest or Cicero is in fact Tully. Even now, we could conceivably discover that we were wrong in supposing that Hesperus was Phosphorus. Maybe the astronomers made an error. So it seems that this view is wrong and that if by a name we do not mean some artificial notion of names such as Russell's, but a proper name in the ordinary sense, then there can be contingent identity statements using proper names, and the view to the contrary seems plainly wrong.

In recent philosophy a large number of other identity statements have been emphasized as examples of contingent identity statements, different, perhaps, from either of the types I have mentioned before. One of them is, for example, the statement "Heat is the motion of molecules." First, science is supposed to have discovered this. Empirical scientists in their investigations have been supposed to discover (and, I suppose, they did) that the external phenomenon which we call "heat" is, in fact, molecular agitation. Another example of such a discovery is that water is H_2O, and yet other examples are that gold is the element with such and such an atomic number, that light is a stream of photons, and so on. These are all in some sense of "identity statement" identity statements. Second, it is thought, they are plainly contingent identity statements, just because they were scientific discoveries. After all, heat might have turned out not to have been the motion of molecules. There were other alternative theories of heat proposed, for example, the caloric theory of heat. If these theories of heat had been correct, then heat would not have been the motion of molecules, but instead, some substance suffusing the hot object, called "caloric". And it was a matter of course of science and not of any logical necessity that the one theory turned out to be correct and the other theory turned out to be incorrect.

So, here again, we have, apparently, another plain example of a contingent identity statement. This has been supposed to be a very important example because of its connection with the mind–body problem. There have been many philosophers who have wanted to be materialists, and to be materialists in a particular form, which is known today as "the identity theory". According to this theory, a certain mental state, such as a person's being in pain, is identical with a certain state of his brain (or, perhaps, of his entire body, according to some theorists), at any rate, a certain material or neural state of his brain or body. And so, according to this theory, my being in pain at this instant, if I were, would be identical with my body's being or my brain's being in a certain state. Others have objected that this cannot be because, after all, we can imagine my pain existing even if the state of the body did not. We can perhaps imagine my not being embodied at all and still being in pain, or, conversely, we could imagine my body existing and being in the very same state even if there were no pain. In fact, conceivably, it could be in this state even though there were no mind 'back of it', so to speak, at all. The usual reply has been to concede that all of these things might have been the case, but to argue that these are irrelevant to the question of the identity of the mental state and the physical state. This identity, it is said, is just another contingent scientific identification, similar to the identification of heat with molecular motion, or water with H_2O. Just as we can imagine heat without any molecular motion, so we can imagine a mental state without any corresponding brain state. But, just as the first fact is not damaging to the identification of heat and the motion of molecules, so the second fact is not at all damaging to the identification of a mental state with the corresponding brain state. And so, many recent philosophers have held it to be very important for our theoretical understanding of the mind–body problem that there can be contingent identity statements of this form.

To state finally what *I* think, as opposed to what seems to be the case, or what others think, I think that in both cases, the case of names and the case of the theoretical identifications, the identity statements are necessary and not contingent. That is to say, they are necessary if *true*; of course, false identity statements are not necessary. How can one possibly defend such a view? Perhaps I lack a complete answer to this question, even though I am convinced that the view is true. But to begin an answer, let me make some distinctions that I want to use. The first is between a *rigid* and a *nonrigid designator*. What do these terms mean? As an example of a nonrigid designator, I can give an expression such as 'the inventor of bifocals'. Let us suppose it was Benjamin Franklin who invented bifocals, and so the expression, 'the inventor of bifocals', designates or refers to a certain man, namely, Benjamin Franklin. However, we can easily imagine that the world could have been different, that under different circumstances someone else would have come upon this invention before Benjamin Franklin did, and in that case, *he* would have been the inventor of bifocals. So, in this sense, the expression 'the inventor of bifocals' is nonrigid: under certain circumstances one man would have been the inventor of bifocals; under other circumstances, another man would have. In contrast, consider the expression 'the square root of 25'. Independently of the empirical facts, we can give an arithmetical proof that the square root of 25 is in fact the number 5, and because we have proved this mathematically, what we have proved is necessary. If we think of numbers as entities at all, and let us suppose, at least for the purpose of this lecture, that we do, then the expression 'the square root of 25' necessarily designates a certain number, namely 5. Such an expression I call 'a *rigid* designator'. Some philosophers think that anyone who even uses the notions of rigid or nonrigid designator has already shown that he has fallen into a certain confusion or has not paid attention to certain facts. What do I mean by 'rigid designator'? I mean a term that designates the same object in all possible worlds. To get rid of one confusion which certainly is not mine, I do not use "might have designated a different object" to refer to the fact that language might have been used differently. For example, the expression 'the inventor of bifocals' might have been used by inhabitants of this planet always to refer to the man who corrupted Hadleyburg. This would have been the case, if, first, the people on this planet had not spoken English, but some other language, which phonetically overlapped with English; and if, second, in that language the expression 'the inventor of bifocals' meant the 'man who corrupted Hadleyburg'. Then it would refer, of course, in their language, to whoever in fact corrupted Hadleyburg in this counterfactual situation. That is not what I mean. What I mean by saying that a description might have referred to something different, I mean that in *our* language as *we* use it in describing a counterfactual situation, there might have been a different object satisfying the descriptive conditions *we* give for reference. So, for example, we use the phrase 'the inventor of bifocals', when we are talking about another possible world or a counterfactual situation, to refer to whoever in that counterfactual situation would have invented bifocals, not to the person whom people *in* that counterfactual situation would have called 'the inventor of bifocals'. *They* might have spoken a different language which phonetically overlapped with English in which 'the inventor of bifocals' is used in some other way. I am *not* concerned with that question here. For that matter, they might have been deaf and dumb, or there might have been no people at all. (There still could have been an inventor of bifocals even if there were no people—God, or Satan, will do.)

Second, in talking about the notion of a rigid designator, I do not mean to imply that the object referred to has to exist in all possible worlds, that is, that it has to necessarily exist. Some things, perhaps mathematical entities such as the positive integers, if they exist at all, necessarily exist. Some people have held that God both exists and necessarily exists; others, that He contingently exists; others, that He contingently fails to exist; and others, that He necessarily fails to exist:[8] all four options have been tried. But at any rate, when I use the notion of rigid designator, I do not imply that the object referred to necessarily exists. All I mean is that in any possible world where the object in question *does* exist, in any situation where the object *would* exist, we use the designator in question to designate that object. In a situation where the object does not exist, then we should say that the designator has no referent and that the object in question so designated does not exist.

As I said, many philosophers would find the very notion of rigid designator objectionable *per se*. And the objection that people make may be stated as follows: look, you're talking about situations which are counterfactual, that is to say, you're talking about other possible worlds. Now these worlds are completely disjoint, after all, from the actual world which is not just another possible world; it is the actual world. So, before you talk about, let us say, such an object as Richard Nixon in another possible world at all, you have to say which object in this other possible world would *be* Richard Nixon. Let us talk about a situation in which, as *you* would say, Richard Nixon would have been a member of SDS.[9] Certainly the member of SDS you are talking about is someone very different in many of his properties from Nixon. Before we even can say whether this man would have been Richard Nixon or not, we have to set up criteria of identity across possible worlds. Here are these other possible worlds. There are all kinds of objects in them with different properties from those of any actual object. Some of them resemble Nixon in some ways, some of them resemble Nixon in other ways. Well, which of these objects is Nixon? One has to give a criterion of identity. And this shows how the very notion of rigid designator runs in a circle. Suppose we designate a certain number as the number of planets. Then, if that is our favourite way, so to speak, of designating this number, then in any other possible worlds we will have to identify whatever number is the number of planets with the number 9, which in the actual world is the number of planets. So, it is argued by various philosophers, for example, implicitly by Quine, and explicitly by many others in his wake, we cannot really ask whether a designator is rigid or nonrigid because we first need a criterion of identity across possible worlds. An extreme view has even been held that, since possible worlds are so disjoint from our

[8] If there is no deity, and especially if the non-existence of a deity is *necessary*, it is dubious that we can use "He" to refer to a deity. The use in the text must be taken to be non-literal.

[9] Ed. note: 'SDS' stands for Students for a Democratic Society, a radical political group that opposed the war in Vietnam.

own, we cannot really say that any object in them is the *same* as an object existing now but only that there are some objects which resemble things in the actual world, more or less. We, therefore, should not really speak of what would have been true of Nixon in another possible world but, only of what 'counterparts' (the term which David Lewis uses[10]) of Nixon there would have been. Some people in other possible worlds have dogs whom they call 'Checkers'. Others favour the ABM but do not have any dog called Checkers. There are various people who resemble Nixon more or less, but none of them can really be said to be Nixon; they are only *counterparts* of Nixon, and you choose which one is the best counterpart by noting which resembles Nixon the most closely, according to your favourite criteria. Such views are widespread, both among the defenders of quantified modal logic and among its detractors.

All of this talk seems to me to have taken the metaphor of possible worlds much too seriously in some way. It is as if a 'possible world' were like a foreign country, or distant planet way out there. It is as if we see dimly through a telescope various actors on this distant planet. Actually David Lewis's view seems the most reasonable if one takes this picture literally. No one far away on another planet can be strictly identical with someone here. But, even if we have some marvellous methods of transportation to take one and the same person from planet to planet, we really need some epistemological criteria of identity to be able to say whether someone on this distant planet is the same person as someone here.

All of this seems to me to be a totally misguided way of looking at things. What it amounts to is the view that counterfactual situations have to be described purely qualitatively. So, we cannot say, for example, "If Nixon had only given a sufficient bribe to Senator X, he would have gotten Carswell through" because that refers to certain people, Nixon and Carswell, and talks about what things would be true of them in a counterfactual situation. We must say instead "If a man who has a hairline like such and such, and holds such and such political opinions had given a bribe to a man who was a senator and had such and such other qualities, then a man who was a judge in the South and had many other qualities resembling Carswell would have been confirmed." In other words, we must describe counterfactual situations purely qualitatively and then ask the question, "Given that the situation contains people or things with such and such qualities, which of these people is

[10] David K. Lewis, 'Counterpart Theory and Quantified Modal Logic', *Journal of Philosophy*, 65 (1968), 113 ff.

(or is a counterpart of) Nixon, which is Carswell, and so on?" This seems to me to be wrong. Who is to prevent us from saying "Nixon might have gotten Carswell through had he done certain things"? We are speaking of *Nixon* and asking what, in certain counterfactual situations, would have been true of *him*. We can say that if Nixon had done such and such, he would have lost the election to Humphrey. Those I am opposing would argue, "Yes, but how do you find out if the man you are talking about is in fact Nixon?" It would indeed be very hard to find out, if you were looking at the whole situation through a telescope, but that is not what we are doing here. Possible worlds are not something to which an epistemological question like this applies. And if the phrase 'possible worlds' is what makes anyone think some such question applies, he should just *drop* this phrase and use some other expression, say 'counterfactual situation', which might be less misleading. If we say "If Nixon had bribed such and such a Senator, Nixon would have gotten Carswell through", what is *given* in the very description of that situation is that it is a situation in which we are speaking of Nixon, and of Carswell, and of such and such a Senator. And there seems to be no less objection to *stipulating* that we are speaking of certain *people* than there can be objection to stipulating that we are speaking of certain *qualities*. Advocates of the other view take speaking of certain qualities as unobjectionable. They do not say, "How do we know that this quality (in another possible world) is that of redness?" But they do find speaking of certain *people* objectionable. But I see no more reason to object in the one case than in the other. I think it really comes from the idea of possible worlds as existing out there, but very far off, viewable only through a special telescope. Even more objectionable is the view of David Lewis. According to Lewis, when we say "Under certain circumstances Nixon would have gotten Carswell through", we really mean "Some man, other than Nixon but closely resembling him, would have gotten some judge, other than Carswell but closely resembling him, through." Maybe that is so, that some man closely resembling Nixon could have gotten some man closely resembling Carswell through. But *that* would not comfort either Nixon or Carswell, nor would it make Nixon kick himself and say "*I* should have done such and such to get Carswell through." The question is whether under certain circumstances Nixon *himself* could have gotten *Carswell* through. And I think the objection is simply based on a misguided picture.

Instead, we can perfectly well talk about rigid and nonrigid designators. Moreover, we have a simple, intuitive test for them. We can say, for example, that the number of planets might have been a different number from the number it in fact is. For example, there might have been only seven planets. We can say that the inventor of bifocals might have been someone other than the man who *in fact* invented bifocals.[11] We cannot say, though, that the square root of 81 might have been a different number from the number it in fact is, for that number just has to be 9. If we apply this intuitive test to proper names, such as for example 'Richard Nixon', they would seem intuitively to come out to be rigid designators. First, when we talk even about the counterfactual situation in which we suppose Nixon to have done different things, we assume we are still talking about Nixon himself. We say, "If Nixon had bribed a certain Senator, he would have gotten Carswell through", and we assume that by 'Nixon' and 'Carswell' we are still referring to the very same people as in the actual world. And it seems that we cannot say "Nixon might have been a different man from the man he in fact was", unless, of course, we mean it metaphorically: He might have been a different *sort* of person (if you believe in free will and that people are not inherently corrupt). You might think the statement true in that sense, but Nixon could not have been in the other literal sense a different person from the person he, in fact, is, even though the thirty-seventh President of the United States might have been Humphrey. So the phrase "the thirty-seventh President" is nonrigid, but 'Nixon', it would seem, is rigid.

Let me make another distinction before I go back to the question of identity statements. This distinction is very fundamental and also hard to see through. In recent discussion, many philosophers who have debated the meaningfulness of various categories of truths, have regarded them as identical. Some of those who identify them are vociferous defenders of them, and others, such as Quine, say they are all identically meaning-

[11] Some philosophers think that definite descriptions, in English, are ambiguous, that sometimes 'the inventor of bifocals' rigidly designates the man who in fact invented bifocals. I am tentatively inclined to reject this view, construed as a thesis about English (as opposed to a possible hypothetical language), but I will not argue the question here.

What I do wish to note is that, contrary to some opinions, this alleged ambiguity cannot replace the Russellian notion of the scope of a description. Consider the sentence, "The number of planets might have been necessarily even." This sentence plainly can be read so as to express a truth; had there been eight planets, the number of planets would have been necessarily even. Yet without scope distinctions, both a 'referential' (rigid) and a non-rigid reading of the description will make the statement false. (Since the number of planets is nine, the rigid reading amounts to the falsity that nine might have been necessarily even.)

The 'rigid' reading is equivalent to the Russellian primary occurrence; the non-rigid, to innermost scope—some, following Donnellan, perhaps loosely, have called this reading the 'attributive' use. The possibility of intermediate scopes is then ignored. In the present instance, the intended reading of $\Diamond\Box$(the number of planets is even) makes the scope of the description $\Box$(the number of planets is even), neither the largest nor the smallest possible.

less. But usually they're not distinguished. These are categories such as 'analytic', 'necessary', 'a priori', and sometimes even 'certain'. I will not talk about all of these but only about the notions of a prioricity and necessity. Very often these are held to be synonyms. (Many philosophers probably should not be described as holding them to be synonyms; they simply *use* them interchangeably.) I wish to distinguish them. What do we mean by calling a statement *necessary*? We simply mean that the statement in question, first, is true, and, second, that it could not have been otherwise. When we say that something is *contingently* true, we mean that, though it is in fact the case, it could have been the case that things would have been otherwise. If we wish to assign this distinction to a branch of philosophy, we should assign it to metaphysics. To the contrary, there is the notion of an *a priori truth*. An a priori truth is supposed to be one which can be *known* to be true independently of all experience. Notice that this does not in and of itself say anything about all possible worlds, unless this is put into the definition. All that it says is that it can be known to be true of the actual world, independently of all experience. It may, by some philosophical argument, follow from our knowing, independently of experience, that something is true of the actual world, that it has to be known to be true also of all possible worlds. But if this is to be established, it requires some philosophical argument to establish it. Now, *this* notion, if we were to assign it to a branch of philosophy, belongs, not to metaphysics, but to epistemology. It has to do with the way we can know certain things to be in fact true. Now, it may be the case, of course, that anything which is necessary is something which *can* be known a priori. (Notice, by the way, the notion a priori truth as thus defined has in it *another* modality: it *can* be known independently of all experience. It is a little complicated because there is a double modality here.) I will not have time to explore these notions in full detail here, but one thing we can see from the outset is that these two notions are by no means trivially the same. If they are coextensive, it takes some philosophical argument to establish it. As stated, they belong to different domains of philosophy. One of them has something to do with *knowledge*, of what can be known in certain ways about the *actual* world. The other one has to do with *metaphysics*, how the world *could* have been; given that it is the way it is, could it have been otherwise, in certain ways? Now I hold, as a matter of fact, that neither class of statements is contained in the other. But, all we need to talk about here is this: Is everything that is necessary knowable a priori or known a priori? Consider the following example: the Goldbach conjecture. This says that every even number is the sum of two primes. It is a mathematical statement and if it is true at all, it has to be necessary. Certainly, one could not say that though in fact every even number is the sum of two primes, there could have been some extra number which was even and not the sum of two primes. What would that mean? On the other hand, the answer to the question whether every even number *is* in fact the sum of two primes is unknown, and we have no method at present for deciding. So we certainly do not know, a priori or even a posteriori, that every even number is the sum of two primes. (Well, perhaps we have some evidence in that no counterexample has been found.) But we certainly do not know a priori anyway, that every even number is, in fact, the sum of two primes. But, of course, the definition just says "*can* be known independently of experience", and someone might say that if it is true, we *could* know it independently of experience. It is hard to see exactly what this claim means. It might be so. One thing it might mean is that if it were true we could *prove* it. This claim is certainly wrong if it is generally applied to mathematical statements and we have to work within some fixed system. This is what Gödel proved. And even if we mean an 'intuitive proof in general' it might just be the case (at least, this view is as clear and as probable as the contrary) that though the statement is true, there is just no way the human mind could ever prove it. Of course, one way an *infinite* mind might be able to prove it is by looking through each natural number one by one and checking. In this sense, of course, it can, perhaps, be known a priori, but only by an infinite mind, and then this gets into other complicated questions. I do not want to discuss questions about the conceivability of performing an infinite number of acts like looking through each number one by one. A vast philosophical literature has been written on this: some have declared it is logically impossible; others that it is logically possible; and some do not know. The main point is that it is not trivial that just because such a statement is necessary it can be known a priori. Some considerable clarification is required before we decide that it can be so known. And so this shows that even if everything necessary is a priori in some sense, it should not be taken as a trivial matter of definition. It is a substantive philosophical thesis which requires some work.

Another example that one might give relates to the problem of essentialism. Here is a lectern. A question which has often been raised in philosophy is: What are its essential properties? What properties, aside from trivial ones like self-identity, are such that this object has to have them if it exists at all,[12] are such that if an object did not have it, it

[12] This definition is the usual formulation of the notion of essential property, but an exception must be made for existence itself: on the definition given, existence would be

would not be this object?[13] For example, being made of wood, and not of ice, might be an essential property of this lectern. Let us just take the weaker statement that it is not made of ice. That will establish it as strongly as we need it, perhaps as dramatically. Supposing this lectern is in fact made of wood, could this very lectern have been made from the very beginning of its existence from ice, say frozen from water in the Thames? One has a considerable feeling that it could *not*, though in fact one certainly could have made a lectern of water from the Thames, frozen it into ice by some process, and put it right there in place of this thing. If one had done so, one would have made, of course, a *different* object. It would not have been *this very lectern*, and so one would not have a case in which this very lectern here was made of ice, or was made from water from the Thames. The question of whether it could afterward, say in a minute from now, turn into ice is something else. So, it would seem, if an example like this is correct—and this is what advocates of essentialism have held—that this lectern could not have been made of ice, that is in any counterfactual situation of which we would say that this lectern existed at all, we would have to say also that it was not made from water from the Thames frozen into ice. Some have rejected, of course, any such notion of essential property as meaningless. Usually, it is because (and I think this is what Quine, for example, would say) they have held that it depends on the notion of identity across possible worlds, and that this is itself meaningless. Since I have rejected this view already, I will not deal with it again. We can talk about *this very object*, and whether it could have had certain properties which it does not in fact have. For example, it could have been in another room from the room it in fact is in, even at this very time, but it could not have been made from the very beginning from water frozen into ice.

---

trivially essential. We should regard existence as essential to an object only if the object necessarily exists. Perhaps there are other *recherché* properties, involving existence, for which the definition is similarly objectionable. (I thank Michael Slote for this observation.)

[13] The two clauses of the sentence footnoted give equivalent definitions of the notion of essential property, since $\Box((\exists x)(x = a) \supset Fa)$ is equivalent to $\Box(x)(\sim Fx \supset x = a)$. The second formulation, however, has served as a powerful seducer in favour of theories of 'identification across possible worlds'. For it suggests that we consider 'an object *b* in another possible world' and test whether it is identifiable with *a* by asking whether it lacks any of the essential properties of *a*. Let me therefore emphasize that, although an essential property is (trivially) a property without which an object cannot be *a*, it by no means follows that the essential, purely qualitative properties of *a* jointly form a sufficient condition for being *a*, nor that *any* purely qualitative conditions are sufficient for an object to be *a*. Further, even if necessary and sufficient qualitative conditions for an object to be Nixon may exist, there would still be little justification for the demand for a purely qualitative description of all counterfactual situations. We can ask whether Nixon might have been a Democrat without engaging in these subtleties.

If the essentialist view is correct, it can only be correct if we sharply distinguish between the notions of a posteriori and a priori truth on the one hand, and contingent and necessary truth on the other hand, for although the statement that this table, if it exists at all, was not made of ice, is necessary, it certainly is not something that we know a priori. What we know is that first, lecterns usually are not made of ice, they are usually made of wood. This looks like wood. It does not feel cold and it probably would if it were made of ice. Therefore, I conclude, probably this is not made of ice. Here my entire judgement is a posteriori. I could find out that an ingenious trick has been played upon me and that, in fact, this lectern is made of ice; but what I am saying is, given that it is in fact not made of ice, in fact is made of wood, one cannot imagine that under certain circumstances it could have been made of ice. So we have to say that though we cannot know a priori whether this table was made of ice or not, given that it is not made of ice, it is *necessarily* not made of ice. In other words, if P is the statement that the lectern is not made of ice, one knows by a priori philosophical analysis, some conditional of the form "if P, then necessarily P". If the table is not made of ice, it is necessarily not made of ice. On the other hand, then, we know by empirical investigation that P, the antecedent of the conditional, is true—that this table is not made of ice. We can conclude by *modus ponens*:

$$\frac{\begin{array}{l} P \supset \Box P \\ P \end{array}}{\Box P}$$

The conclusion—'$\Box P$'—is that it is necessary that the table not be made of ice, and this conclusion is known a posteriori, since one of the premises on which it is based is a posteriori. So, the notion of essential properties can be maintained only by distinguishing between the notions of a priori and necessary truth, and I do maintain it.

Let us return to the question of identities. Concerning the statement 'Hesperus is Phosphorus' or the statement 'Cicero is Tully', one can find all of these out by empirical investigation, and we might turn out to be wrong in our empirical beliefs. So, it is usually argued, such statements must therefore be contingent. Some have embraced the other side of the coin and have held "Because of this argument about necessity, identity statements between names have to be knowable a priori, so, only a very special category of names, possibly, really works as names; the other things are bogus names, disguised descriptions, or something of the sort.

However, a certain very narrow class of statements of identity are known a priori, and these are the ones which contain the genuine names." If one accepts the distinctions that I have made, one need not jump to either conclusion. One can hold that certain statements of identity between names, though often known a posteriori, and maybe not knowable a priori, are in fact necessary, if true. So, we have some room to hold this. But, of course, to have some room to hold it does not mean that we should hold it. So let us see what the evidence is. First, recall the remark that I made that proper names seem to be rigid designators, as when we use the name 'Nixon' to talk about a certain man, even in counterfactual situations. If we say, "If Nixon had not written the letter to Saxbe, maybe he would have gotten Carswell through", we are in this statement talking about Nixon, Saxbe, and Carswell, the very same men as in the actual world, and what would have happened to them under certain counterfactual circumstances. If names are rigid designators, then there can be no question about identities being necessary, because 'a' and 'b' will be rigid designators of a certain man or thing x. Then even in every possible world, a and b will both refer to this same object x, and to no other, and so there will be no situation in which a might not have been b. That would have to be a situation in which the object which we are also now calling 'x' would not have been identical with itself. Then one could not possibly have a situation in which Cicero would not have been Tully or Hesperus would not have been Phosphorus.[14]

Aside from the identification of necessity with a priority, what has made people feel the other way? There are two things which have made people feel the other way.[15] Some people tend to regard identity statements as metalinguistic statements, to identify the statement "Hesperus is Phosphorus" with the metalinguistic statement, "'Hesperus' and

[14] I thus agree with Quine, that "Hesperus is Phosphorus" is (or can be) an empirical discovery; with Marcus, that it is necessary. Both Quine and Marcus, according to the present standpoint, err in identifying the epistemological and the metaphysical issues.

[15] The two confusions alleged, especially the second, are both related to the confusion of the metaphysical question of the necessity of "Hesperus is Phosphorus" with the epistemological question of its a prioricity. For if Hesperus is identified by its position in the sky in the evening, and Phosphorus by its position in the morning, an investigator may well know, in advance of empirical research, that Hesperus is Phosphorus if and only if one and the same body occupies position x in the evening and position y in the morning. The a priori material equivalence of the two statements, however, does not imply their strict (necessary) equivalence. (The same remarks apply to the case of heat and molecular motion below.) Similar remarks apply to some extent to the relationship between "Hesperus is Phosphorus" and "'Hesperus' and 'Phosphorus' name the same thing." A confusion that also operates is, of course, the confusion between what *we* say of a counterfactual situation and how people *in* that situation would have described it; this confusion, too, is probably related to the confusion between a prioricity and necessity.

'Phosphorus' are names of the same heavenly body." And that, of course, might have been false. We might have used the terms 'Hesperus' and 'Phosphorus' as names of *two* different heavenly bodies. But, of course, this has nothing to do with the necessity of identity. In the same sense "2 + 2 = 4" might have been false. The phrases "2 + 2" and "4" might have been used to refer to two different numbers. One can imagine a language, for example, in which "+", "2", and "=" were used in the standard way, but "4" was used as the name of, say, the square root of minus 1, as we should call it, "*i*". Then "2 + 2 = 4" would be false, for 2 plus 2 is not equal to the square root of minus 1. But this is not what we want. We do not want just to say that a certain statement which we in fact use to express something true could have expressed something false. We want to use the statement in *our* way and see if it could have been false. Let us do this. What is the idea people have? They say, "Look, Hesperus might not have been Phosphorus. Here a certain planet was seen in the morning, and it was seen in the evening; and it just turned out later on as a matter of empirical fact that they were one and the same planet. If things had turned out otherwise, they would have been two different planets, or two different heavenly bodies, so how can you say that such a statement is necessary?"

Now there are two things that such people can mean. First, they can mean that we do not know a priori whether Hesperus is Phosphorus. This I have already conceded. Second, they may mean that they can actually imagine circumstances that they would call circumstances in which Hesperus would not have been Phosphorus. Let us think what would be such a circumstance, using these terms here as *names* of a planet. For example, it could have been the case that Venus did indeed rise in the morning in exactly the position in which we saw it, but that on the other hand, in the position which is in fact occupied by Venus in the evening, Venus was not there, and Mars took its place. This is all counterfactual because in fact Venus is there. Now one can also imagine that in this counterfactual other possible world, the earth would have been inhabited by people and that they should have used the names 'Phosphorus' for Venus in the morning and 'Hesperus' for Mars in the evening. Now, this is all very good, but would it be a situation in which Hesperus was not Phosphorus? Of course, it is a situation in which people would have been able to *say*, truly, "Hesperus is not Phosphorus"; but we are supposed to describe things in our language, not in theirs. So let us describe it in our language. Well, how could it actually happen that Venus would not be in that position in the evening? For example, let us say that there is some comet that comes around every evening and yanks things over a little bit. (That would be a very simple scientific way of imagining it: not really too simple—that is very hard to imagine actually.) It just happens to come around every evening, and things get yanked over a bit. Mars gets yanked over to the very position where Venus is, then the comet yanks things back to their normal position in the morning. Thinking of this planet which we now call 'Phosphorus', what should we say? Well, we can say that the comet passes it and yanks Phosphorus over so that it is not in the position normally occupied by Phosphorus in the evening. If we do say this, and really use 'Phosphorus' as the name of a planet, then we have to say that, under such circumstances, Phosphorus in the evening would not be in the position where we, in fact, saw it; or alternatively, Hesperus in the evening would not be in the position in which we, in fact, saw it. We might say that under such circumstances, we would not have called Hesperus 'Hesperus' because Hesperus would have been in a different position. But that still would not make Phosphorus different from Hesperus; but what would then be the case instead is that Hesperus would have been in a different position from the position it in fact is and, perhaps, not in such a position that people would have called it 'Hesperus'. But that would not be a situation in which Phosphorus would not have been Hesperus.

Let us take another example which may be clearer. Suppose someone uses 'Tully' to refer to the Roman orator who denounced Cataline and uses the name 'Cicero' to refer to the man whose works he had to study in third-year Latin in high school. Of course, he may not know in advance that the very same man who denounced Cataline wrote these works, and that is a contingent statement. But the fact that this statement is contingent should not make us think that the statement that Cicero is Tully, if it is true, and it is in fact true, is contingent. Suppose, for example, that Cicero actually did denounce Cataline, but thought that this political achievement was so great that he should not bother writing any literary works. Would we say that these would be circumstances under which he would not have been Cicero? It seems to me that the answer is no, that instead we would say that, under such circumstances, Cicero would not have written any literary works. It is not a necessary property of Cicero—the way the shadow follows the man—that he should have written certain works; we can easily imagine a situation in which Shakespeare would not have written the works of Shakespeare, or one in which Cicero would not have written the works of Cicero. What may be the case is that we *fix the reference* of the term 'Cicero' by use of some descriptive phrase, such as 'the author of these works'. But once we have this reference fixed, we then use the name

'Cicero' *rigidly* to designate the man who in fact we have identified by his authorship of these works. We do not use it to designate whoever would have written these works in place of Cicero, if someone else wrote them. It might have been the case that the man who wrote these works was not the man who denounced Cataline. Cassius might have written these works. But we would not then say that Cicero would have been Cassius, unless we were speaking in a very loose and metaphorical way. We would say that Cicero, whom we may have identified and come to know by his works, would not have written them, and that someone else, say Cassius, would have written them in his place.

Such examples are not grounds for thinking that identity statements are contingent. To take them as such grounds is to misconstrue the relation between a *name* and a *description used to fix its reference*, to take them to be *synonyms*. Even if we fix the reference of such a name as 'Cicero' as the man who wrote such and such works, in speaking of counterfactual situations, when we speak of Cicero, we do not then speak of whoever in such counterfactual situations *would* have written such and such works, but rather of Cicero, whom we have identified by the contingent property that he is the man who in fact, that is, in the actual world, wrote certain works.[16]

I hope this is reasonably clear in a brief compass. Now, actually I have been presupposing something I do not really believe to be, in general, true. Let us suppose that we do fix the reference of a name by a description. Even if we do so, we do not then make the name *synonymous* with the description, but instead we use the name *rigidly* to refer to the object so named, even in talking about counterfactual situations where the thing named would not satisfy the description in question. Now, this is what I think in fact is true for those cases of naming where the reference is fixed by description. But, in fact, I also

[16] If someone protests, regarding the lectern, that it *could* after all have *turned out* to have been made of ice, and therefore could have been made of ice, I would reply that what he really means is that *a lectern* could have looked just like this one, and have been placed in the same position as this one, and yet have been made of ice. In short, I could have been in the *same epistemological situation* in relation to *a lectern made of ice* as I actually am in relation to *this* lectern. In the main text, I have argued that the same reply should be given to protests that Hesperus could have turned out to be other than Phosphorus, or Cicero other than Tully. Here, then, the notion of 'counterpart' comes into its own. For it is not this table, but an epistemic 'counterpart', which was hewn from ice; not Hesperus–Phosphorus–Venus, but two distinct counterparts thereof, in two of the roles Venus actually plays (that of Evening Star and Morning Star), which are different. Precisely because of this fact, it is not *this table* which could have been made of ice. Statements about the modal properties of *this table* never refer to counterparts. However, if someone confuses the epistemological and the metaphysical problems, he will be well on the way to the counterpart theory Lewis and others have advocated.

think, contrary to most recent theorists, that the reference of names is rarely or almost never fixed by means of description. And by this I do not just mean what Searle says: "It's not a single description, but rather a cluster, a family of properties which fixes the reference." I mean that properties in this sense are not used *at all*. But I do not have the time to go into this here. So, let us suppose that at least one half of prevailing views about naming is true, that the reference is fixed by descriptions. Even were that true, the name would not be synonymous with the description, but would be used to *name* an object which we pick out by the contingent fact that it satisfies a certain description. And so, even though we can imagine a case where the man who wrote these works would not have been the man who denounced Cataline, we should not say that that would be a case in which Cicero would not have been Tully. We should say that it is a case in which Cicero did not write these works, but rather that Cassius did. And the identity of Cicero and Tully still holds.

Let me turn to the case of heat and the motion of molecules. Here surely is a case that is contingent identity! Recent philosophy has emphasized this again and again. So, if it is a case of contingent identity, then let us imagine under what circumstances it would be false. Now, concerning this statement I hold that the circumstances philosophers apparently have in mind as circumstances under which it would have been false are not in fact such circumstances. First, of course, it is argued that "Heat is the motion of molecules" is an a posteriori judgement; scientific investigation might have turned out otherwise. As I said before, this shows nothing against the view that it is necessary—at least if I am right. But here, surely, people had very specific circumstances in mind under which, so they thought, the judgement that heat is the motion of molecules would have been false. What were these circumstances? One can distill them out of the fact that we found out empirically that heat is the motion of molecules. How was this? What did we find out first when we found out that heat is the motion of molecules? There is a certain external phenomenon which we can sense by the sense of touch, and it produces a sensation which we call "the sensation of heat". We then discover that the external phenomenon which produces this sensation, which we sense, by means of our sense of touch, is in fact that of molecular agitation in the thing that we touch, a very high degree of molecular agitation. So, it might be thought, to imagine a situation in which heat would not have been the motion of molecules, we need only imagine a situation in which we would have had the very same sensation and it would have been produced by something other than the motion of molecules. Similarly, if we wanted to imagine a situation in which light was not a stream of photons, we could imagine a situation in which we were sensitive to something else in exactly the same way, producing what we call visual experiences, though not through a stream of photons. To make the case stronger, or to look at another side of the coin, we could also consider a situation in which we *are* concerned with the motion of molecules but in which such motion does not give us the sensation of heat. And it might also have happened that we, or, at least, the creatures inhabiting this planet, might have been so constituted that, let us say, an increase in the motion of molecules did not give us this sensation but that, on the contrary, a slowing down of the molecules did give us the very same sensation. This would be a situation, so it might be thought, in which heat would not be the motion of molecules, or, more precisely, in which temperature would not be mean molecular kinetic energy.

But I think it would not be so. Let us think about the situation again. First, let us think about it in the actual world. Imagine right now the world invaded by a number of Martians, who do indeed get the very sensation that we call "the sensation of heat" when they feel some ice which has slow molecular motion, and who do not get a sensation of heat—in fact, maybe just the reverse—when they put their hand near a fire which causes a lot of molecular agitation. Would we say, "Ah, this casts some doubt on heat being the motion of molecules, because there are these other people who don't get the same sensation"? Obviously not, and no one would think so. We would say instead that the Martians somehow feel the very sensation we get when we feel heat when they feel cold and that they do not get a sensation of heat when they feel heat. But now let us think of a counterfactual situation.[17] Suppose the earth had from the very beginning been inhabited by such creatures. First, imagine it inhabited by no creatures at all: then there is no one to feel any sensations of heat. But we would not say that under such circumstances it would necessarily be the case that heat did not exist; we would say that heat might have existed, for example, if there were fires that heated up the air.

[17] Isn't the situation I just described also counterfactual? At least it may well be, if such Martians never in fact invade. Strictly speaking, the distinction I wish to draw compares how we *would* speak *in* a (possibly counterfactual) situation, *if* it obtained, and how we *do* speak *of* a counterfactual situation, knowing that it does not obtain—i.e., the distinction between the language we would have used in a situation and the language we *do* use to describe it. (Consider the description: "Suppose we all spoke German." This description is in English.) The former case can be made vivid by imagining the counterfactual situation to be actual.

Let us suppose the laws of physics were not very different: Fires do heat up the air. Then there would have been heat even though there were no creatures around to feel it. Now let us suppose evolution takes place, and life is created, and there are some creatures around. But they are not like us, they are more like the Martians. Now would we say that heat has suddenly turned to cold, because of the way the creatures of this planet sense it? No, I think we should describe this situation as a situation in which, though the creatures on this planet got our sensation of heat, they did not get it when they were exposed to heat. They got it when they were exposed to cold. And that is something we can surely well imagine. We can imagine it just as we can imagine our planet being invaded by creatures of this sort. Think of it in two steps. First there is a stage where there are no creatures at all, and one can certainly imagine the planet still having both heat and cold, though no one is around to sense it. Then the planet comes through an evolutionary process to be peopled with beings of different neural structure from ourselves. Then these creatures could be such that they were insensitive to heat; they did not feel it in the way we do; but on the other hand, they felt cold in much the same way that we feel heat. But still, heat would be heat, and cold would be cold. And particularly, then, this goes in no way against saying that in this counterfactual situation heat would still *be* the molecular motion, *be* that which is produced by fires, and so on, just as it would have been if there had been no creatures on the planet at all. Similarly, we could imagine that the planet was inhabited by creatures who got visual sensations when there were sound waves in the air. We should not therefore say, "Under such circumstances, sound would have been light." Instead we should say, "The planet was inhabited by creatures who were in some sense visually sensitive to sound, and maybe even visually sensitive to light." If this is correct, it can still be and will still be a necessary truth that heat is the motion of molecules and that light is a stream of photons.

To state the view succinctly: we use both the terms 'heat' and 'the motion of molecules' as rigid designators for a certain external phenomenon. Since heat is in fact the motion of molecules, and the designators are rigid, by the argument I have given here, it is going to be *necessary* that heat is the motion of molecules. What gives us the illusion of contingency is the fact we have identified the heat by the contingent fact that there happen to be creatures on this planet—(namely, ourselves) who are sensitive to it in a certain way, that is, who are sensitive to the motion of molecules or to heat—these are one and the same thing. And this is contingent. So we use the description, 'that which causes such and such sensations, or that which we sense in such and such a way', to identify heat. But in using this fact we use a contingent property of heat, just as we use the contingent property of Cicero as having written such and such works to identify him. We then use the terms 'heat' in the one case and 'Cicero' in the other *rigidly* to designate the objects for which they stand. And of course the term 'the motion of molecules' is rigid; it always stands for the motion of molecules, never for any other phenomenon. So, as Bishop Butler said, "everything is what it is and not another thing." Therefore, "Heat is the motion of molecules" will be necessary, not contingent, and one only has the *illusion* of contingency in the way one could have the illusion of contingency in thinking that this table might have been made of ice. We might think one could imagine it, but if we try, we can see on reflection that what we are really imagining is just there being another lectern in this very position here which was in fact made of ice. The fact that we may identify this lectern by being the object we see and touch in such and such a position is something else.

Now how does this relate to the problem of mind and body? It is usually held that this is a contingent identity statement just like "Heat is the motion of molecules." That cannot be. It cannot be a contingent identity statement just like "Heat is the motion of molecules" because, if I am right, "Heat is the motion of molecules" is not a contingent identity statement. Let us look at this statement. For example, "My being in pain at such and such a time is my being in such and such a brain state at such and such a time", or, "Pain in general is such and such a neural (brain) state."

This is held to be contingent on the following grounds. First, we can imagine the brain state existing though there is no pain at all. It is only a scientific fact that whenever we are in a certain brain state we have a pain. Second, one might imagine a creature being in pain, but not being in any specified brain state at all, maybe not having a brain at all. People even think, at least prima facie, though they may be wrong, that they can imagine totally disembodied creatures, at any rate certainly not creatures with bodies anything like our own. So it seems that we can imagine definite circumstances under which this relationship would have been false. Now, if these circumstances are circumstances, notice that we cannot deal with them simply by saying that this is just an illusion, something we can apparently imagine, but in fact cannot in the way we thought erroneously that we could imagine a situation in which heat was not the motion of molecules. Because although we can say that we pick out heat contingently by the contingent property that it affects us in such and such a way, we cannot similarly say that we pick out pain contingently by the fact that it affects us in such and such a way. On such a picture there would be the brain state, and we pick it out by the contingent fact that it affects us as pain. Now that might be true of the brain state, but it cannot be true of the pain. The experience itself has to be *this experience*, and I cannot say that it is a contingent property of the pain I now have that it is a pain.[18] In fact, it would seem that both the terms, 'my pain' and 'my being in such and such a brain state' are, first of all, both rigid designators. That is, whenever anything is such and such a pain, it is essentially that very object, namely, such and such a pain, and wherever anything is such and such a brain state, it is essentially that very object, namely, such and such a brain state. So both of these are rigid designators. One cannot say this pain might have been something else, some other state. These are both rigid designators.

Second, the way we would think of picking them out—namely, the pain by its being an experience of a certain sort, and the brain state by its being the state of a certain material object, being of such and such molecular configuration—both of these pick out their objects essentially and not accidentally, that is, they pick them out by essential properties. Whenever the molecules *are* in this configuration, we *do* have such and such a brain state. Whenever you feel *this*, you do have a pain. So it seems that the identity theorist is in some trouble, for, since we have two rigid designators, the identity statement in question is necessary.

[18] The most popular identity theories advocated today explicitly fail to satisfy this simple requirement. For these theories usually hold that a mental state is a brain state, and that what makes the brain state into a mental state is its 'causal role', the fact that it tends to produce certain behaviour (as intentions produce actions, or pain, pain behaviour) and to be produced by certain stimuli (e.g., pain, by pinpricks). If the relations between the brain state and its causes and effects are regarded as contingent, then *being such-and-such-a-mental-state* is a contingent property of the brain state. Let X be a pain. The causal-role indentity theorist holds (1) that X is a brain state, (2) that the fact that X is a pain is to be analysed (roughly) as the fact that X is produced by certain stimuli and produces certain behaviour. The fact mentioned in (2) is, of course, regarded as contingent; the brain state X might well exist and not tend to produce the appropriate behaviour in the absence of other conditions. Thus (1) and (2) assert that a certain pain X might have existed, yet not have been a pain. This seems to me self-evidently absurd. Imagine any pain: is it possible that *it itself* could have existed, yet not have been a pain?

If $X = Y$, then X and Y share all properties, including modal properties. If X is a pain and Y the corresponding brain state, then *being a pain* is an essential property of X, and *being a brain state* is an essential property of Y. If the correspondence relation is, in fact, identity, then it must be *necessary* of Y that it corresponds to a pain, and *necessary* of X that it correspond to a brain state, indeed to this particular brain state, Y. Both assertions seem false; it *seems* clearly possible that X should have existed without the corresponding brain state; or that the brain state should have existed without being felt as pain. Identity theorists cannot, contrary to their almost universal present practice, accept these intuitions; they must deny them, and explain them away. This is none too easy a thing to do.

Because they pick out their objects essentially, we cannot say the case where you seem to imagine the identity statement false is really an illusion like the illusion one gets in the case of heat and molecular motion, because that illusion depended on the fact that we pick out heat by a certain contingent property. So there is very little room to manoeuvre; perhaps none.[19] The identity theorist, who holds that pain is the brain state, also has to hold that it necessarily is the brain state. He therefore cannot concede, but has to deny, that there would have been situations under which one would have had pain but not the corresponding brain state. Now usually in arguments on the identity theory, this is very far from being denied. In fact, it is conceded from the outset by the materialist as well as by his opponent. He says, "Of course, it *could* have been the case that we had pains without the brain states. It is a contingent identity." But that cannot be. He has to hold that we are under some illusion in thinking that we can imagine that there could have been pains without brain states. And the only model I can think of for what the illusion might be, or at least the model given by the analogy the materialists themselves suggest, namely, heat and molecular motion, simply does not work in this case. So the materialist is up against a very stiff challenge. He has to show that these things we think we can see to be possible are in fact not possible. He has to show that these things which we can imagine are not in fact things we can imagine. And that requires some very different philosophical argument from the sort which has been given in the case of heat and molecular motion. And it would

[19] A brief restatement of the argument may be helpful here. If "pain" and "C-fibre stimulation" are rigid designators of phenomena, one who identifies them must regard the identity as necessary. How can this necessity be reconciled with the apparent fact that C-fibre stimulation might have turned out not to be correlated with pain at all? We might try to reply by analogy to the case of heat and molecular motion; the latter identity, too, is necessary, yet someone may believe that, before scientific investigation showed otherwise, molecular motion might have turned out not to be heat. The reply is, of course, that what really is possible is that people (or some rational or sentient beings) could have been in the *same epistemic situation* as we actually are, and identify *a phenomenon* in the same way we identify heat, namely, by feeling it by the sensation we call "the sensation of heat", without the phenomenon being molecular motion. Further, the beings might not have been sensitive to molecular motion (i.e., to heat) by any neural mechanism whatsoever. It is impossible to explain the apparent possibility of C-fibre stimulations not having been pain in the same way. Here, too, we would have to suppose that we could have been in the same epistemological situation, and identify something in the same way we identify pain, without its corresponding to C-fibre stimulation. But the way we identify pain is by feeling it, and if a C-fibre stimulation could have occurred without our feeling any pain, then the C-fibre stimulation would have occurred without there *being* any pain, contrary to the necessity of the identity. The trouble is that although 'heat' is a rigid designator, heat is picked out by the contingent property of its being felt in a certain way; pain, on the other hand, is picked out by an essential (indeed necessary and sufficient) property. For a sensation to be *felt* as pain is for it to *be* pain.

have to be a deeper and subtler argument than I can fathom and subtler than has ever appeared in any materialist literature that I have read. So the conclusion of this investigation would be that the analytical tools we are using go against the identity thesis and so go against the general thesis that mental states are just physical states.[20]

The next topic would be my own solution to the mind–body problem, but that I do not have.

[20] All arguments against the identity theory which rely on the necessity of identity, or on the notion of essential property, are, of course, inspired by Descartes's argument for his dualism. The earlier arguments which superficially were rebutted by the analogies of heat and molecular motion, and the bifocals inventor who was also Postmaster General, had such an inspiration: and so does my argument here. R. Albritton and M. Slote have informed me that they independently have attempted to give essentialist arguments against the identity theory, and probably others have done so as well.

The simplest Cartesian argument can perhaps be restated as follows: Let 'A' be a *name* (rigid designator) of Descartes's body. Then Descartes argues that since he could exist even if A did not, $\Diamond(\text{Descartes} \neq A)$, hence $\text{Descartes} \neq A$. Those who have accused him of a modal fallacy have forgotten that 'A' is rigid. His argument is valid, and his conclusion is correct, provided its (perhaps dubitable) premiss is accepted. On the other hand, provided that Descartes is regarded as having ceased to exist upon his death, "$\text{Descartes} \neq A$" can be established without the use of a modal argument; for if so, no doubt A survived Descartes when A was a corpse. Thus A had a property (existing at a certain time) which Descartes did not. The same argument can establish that a statue is not the hunk of stone, or the congery, of molecules, of which it is composed. Mere non-identity, then, may be a weak conclusion. (See D. Wiggins, *Philosophical Review*, 77 (1968), 90 ff.) The Cartesian modal argument, however, surely can be deployed to maintain relevant stronger conclusions as well.

XI

PUTNAM'S DOCTRINE OF NATURAL KIND WORDS AND FREGE'S DOCTRINES OF SENSE, REFERENCE, AND EXTENSION: CAN THEY COHERE?

DAVID WIGGINS

1. Hilary Putnam has been apt to emphasize all the differences between the deictic doctrine that he advocates for the understanding of our understanding of natural kind substantives and the accounts of the meanings of these expressions that would have had to be offered by his predecessors in the philosophy of meaning. Delighting in iconoclasm, he has sought at various times to include within the ambit of his entertaining criticisms of his predecessors such figures as Aristotle, the Scholastics, Locke, Mill, Frege, linguistic philosophers, analytical philosophers, philosophers of linguistics, indeed practically everyone.[1]

In this essay, I set out Putnam's proposal and show how, twenty years ago, it broke the mould for one kind of philosophical analysis. But I also try to show that we may deploy Putnam's proposal most convincingly if, flying in the face of Putnam's own wish, we try to place it within the framework of Fregean sense and reference. Then I try to show that, having done that, we can improve our understanding further if we seek to integrate the deictic proposal—significantly but desirably adjusted at one key point—with an extant, neglected but even more time-honoured tradition of semantic speculation, a tradition, not empiricist, in which there is already a clear place for Putnam's insight into the functioning of natural kind words.

Paper delivered at the Conference on the Philosophy of Hilary Putnam, autumn 1990, St Andrews. An earlier version of this paper was given in French at the Institut d'Histoire et Philosophie des Sciences et des Techniques, Paris in 1989, and will be published in *Revue de théologie et de philosophie*, 1992/3.

[1] See for instance the recent exposition that he gives in *Representations and Reality* (Cambridge, Mass.: MIT, 1988).

2. I begin by reminding you of the contents of 'Is Semantics Possible?'. This is a paper that Putnam read to a conference in Brockport, New York, in 1967 (a paper long known only by report and that I came to know myself from the notes taken at the conference by my former student, Ronald de Sousa, to whom I must owe my first appreciation of its significance).[2] It was in this paper that Putnam first introduced the idea that to impart the meaning of a natural kind term is to impart certain core facts: (1) the *stereotype*—consisting of the facts an ordinary speaker needs to know in order to use a natural kind term—and (2) the *extension*, the identification of the latter being the province of experts. Or, as I would rather say (see below, and see my *Sameness and Substance*, Preamble), it was in this paper that Putnam introduced the idea that the sense of thing-kind words standing for natural kinds is reality-invoking or extension-involving.

Putnam led up to his conclusion by criticizing incisively and amusingly the easy (or fall back) supposition that the right way to give the meaning of "lemon", "tiger", "water", or whatever, would have to be by analysis into simpler terms or by giving necessary and sufficient conditions. I worried, when I read de Sousa's notes, that Putnam filed no report of an analytical philosopher (contrast one philosopher of linguistics) with his trousers down actually attempting such a thing. But then, by 1967, linguistic philosophers were a canny crowd; and it quickly appeared that Putnam's failure to identify such an attempt was not a cause to criticize him. The thing that mattered was that the problem of natural kind words was not to be solved by the tact or good taste in which analytical philosophers then excelled, or by the refusal to recognize that these expressions constituted a special question for the philosophy of meaning.

On this matter, as on others, ideas were at that time in short supply. But it must be recorded that, for the slightly similar case of proper names, there had existed from well before 1960 an important minority opinion which (perhaps under the influence of Geach and Anscombe) Michael Dummett took very seriously and which (under Dummett's influence) I myself took maximally seriously (indeed believed and preached to my students from 1959 onwards). This was that the senses of proper names were reality-invoking or object-involving or that a proper name has its meaning, and thus affects the truth-conditions of the sentences in which it occurs, by standing for its bearer: and that there is no other way to give the sense of a proper name than to say which

[2] Eventually this paper was published in H. E. Kiefer and M. K. Munitz (eds.), *Language, Belief and Metaphysics* (Albany, NY: SUNY Press, 1970), 50–63, and in *Metaphilosophy*, 1 (1970), 187–201.

object it is that the name stands for.[3] It was not out of the question then, even in that distant epoch, to say that the senses of proper names were reality-invoking or object-involving or that a proper name had its sense by being assigned to something, not by the laying down of a specification such that a bearer of the name bears that name by virtue of meeting that specification. Nor, among those who held the minority opinion I have just described, was it an unfamiliar question what it would involve to find room for this in Fregean semantics. (The notion of *Art des Gegebenseins* seemed to be ready-made for such an attempt. See Section 8 below.) Rather, that which was still missing in linguistic philosophy and in the philosophy of science of those times was any strong perception of the need to say something similar for potentially predicative expressions like "lemon", "tiger", "water", or to try to *generalize* from the direction of semantical fit that is so strikingly exemplified by the assignment of an ordinary name to something.[4]

Back now to 'Is Semantics Possible?'. Not only, Putnam insisted, were philosophical analyses of substantives such as "lemon", "tiger", "water" laughably inadequate. The only explanation of anyone's even supposing that it might be possible for there to be such an analysis was sheer negligence of the whole *mise-en-scène*, the whole social-cum-technological context, on which we all depend in order to come to understand one another. This context is the thing we depend upon even to understand what we find in a dictionary. Philosophers who were ready to suppose that there could be a philosophical analysis of the meaning of "lemon" or "tiger" were ignoring the division of mental labour and the role of the authorities or experts who sustain our shared understanding of natural kind words. Philosophers should cease to complain, Putnam said, about the fact the dictionaries are "cluttered up with colour samples and stray pieces of empirical information (e.g. the weight of aluminium) not sharply distinguished from purely linguistic information". They should take that fact seriously as a clue to the real situation, the situation that Putnam himself wanted (however schematically) to describe.

3. Confining oneself to the ideas of 'Is Semantics Possible?', but departing a little from his mode of exposition there, one can put the positive

[3] See G. E. M. Anscombe, *An Introduction to Wittgenstein's Tractatus* (London: Hutchinson, 1959), 41, 42, 44; Michael Dummett, 'Truth', *Proc. Ar. Soc.* 59 (1958–9), 141–62.

[4] If you doubt that, read the reply to Putnam in the Brockport volume or ask me for my recollections of the experience of trying to expound Putnam's theory to an incredulous audience at a meeting of the British Association for the Philosophy of Science in the early 1970s. For "direction of fit", see J. L. Austin, 'How to Talk: Some Simple Ways', reprinted in his *Philosophical Papers* (Oxford: Clarendon Press, 1961), and M. J. Woods, as cited in my *Sameness and Substance* (Oxford: Blackwell, 1980).

proposal Putnam wanted to advance as follows: where the instructor's grasp of extension is authoritative (or is downstream from an authoritative identification of that extension), an instructor could initiate a learner into the meaning of the word "lemon" as follows:

> This is a lemon. [Here the instructor displays a specimen.] A thing is a lemon if it resembles this [the specimen] or this [another displayed specimen] or this [a third specimen] in the relevant way. I say *the relevant way*. But to understand better what that way is you must inquire, just to the extent necessary for your purposes, into the *nature* of these three things that I am showing you.

The philosophical claim is that (however artificially) such a demonstration or ostension reconstructs the ordinary teaching and learning of thing-kind words. It reconstructs that which is essential to the transactions that take place between those who know and those who do not know what a given substantive means.

4. I suspect that it is hard for those who have been introduced to philosophy by anyone who understood the point of such a proposal to appreciate the novelty it once enjoyed, at the time when Putnam introduced it. But before I say any more of this novelty or touch upon the tractability of Putnam's doctrine to the theory of sense and reference or trace its affinity to certain rationalist ideas, three points need to be registered.

First, the doctrine, which is sometimes loosely called "the indexical theory", does not, on a true understanding, imply any close similarity between natural-kind substantives and indexicals or demonstratives. If "lemon" or "tiger" or "water" had any real resemblance to "this" or "that" or "now" or "today", these substantives might in other contexts, and without change of lexical meaning, pick out other kinds of things than the kinds that we denominate lemons, tigers, or water. But the point of the theory is to attach the meaning of these words to the real natures, more or less well known, of the actual lemons, tigers, and water that we have encountered. Therefore we must not compare "water" to a demonstrative. The theory is a deictic theory in just one sense: it is the theory of the *deixis* by which we can, under special and favourable conditions, attach a word to a kind of thing.

Secondly, the doctrine does not extend to all thing-kind words. Occasionally, in later papers, Putnam was tempted to apply it to substantives like "pencil". But that was a pity, indeed threatened the shipwreck of a good idea. "Pencil" denominates a functional or instrumental kind.[5] Such a kind might well be defined (even in the strict old-fashioned acceptation of "define"); and this is just as well, because almost the only nomological generalization one will discover by investigation of the class of pencils is the undependable generalization that one can write or draw with a pencil. There is little or no resemblance here to the case where the ostension of a natural kind invites us to extrapolate freely across the observed properties of its exemplars in the search for interesting generalizations about its nature.

This point leads to a third, which might itself have been the occasion for a whole essay, but is not the occasion for the whole of mine. In this area there are problems of the underdetermination of meaning by *deixis* and problems of the proper representativeness of specimens (problems analogous to those that Goodman in *The Structure of Appearance* called "constant companionship" and "imperfect community"). Room must be made within the theory of this demonstrative practice for an instructor to make his pupil understand whether it is a species that is being indicated or a subspecies/variety (is it rose or Rosa Rugosa?); and room must be made for the instructor to make the pupil understand whether something specific or something generic is to be identified (tiger, say, or Felis).[6] Ostension had better not be the magical solution to the problem of natural kind predicates. We shall return to this point in Section 13.

5. So much in outline for the new theory of natural kind words. Now one might ask why it was so difficult for so long for our sort of philosophy to arrive at this deictic or extension-involving conception of their meaning? And how indeed was the barrier (whatever it was) ever surmounted?

Well, I shall leave it to Hilary Putnam to answer the second question—unless modesty forbids. But about the first, there is at least one thing that cries out to be said. This thing is very obvious, but, since it occurred to me altogether afresh the other day, when I was reading a review in *Mind and Language* (Autumn issue, 1990) of the book

[5] In discussion, Hilary Putnam reminded me that at this point he had been reporting a discussion with Rogers Albritton, in which they had been envisaging circumstances under which it was found that all extant pencils had a certain microstructure. But *all* pencils, however manufactured and for whatever specialized purpose? This would be magic. Or (more likely) there would have to have been a practical joke somewhere in this story.

[6] These problems are more tractable than the problems they superficially resemble and that were thrown up by the resemblance (or no-universals) theory of universals defended by Russell and H. H. Price. They are more tractable because here *deixis* can be supported by context and by verbal explanations that are unconstrained by special requirements of ontological parsimony. See the explanations envisaged below, especially Section 13.

Representations and Reality—which is Putnam's recent critique of many positions in present-day philosophy, not least of functionalism—I shall yield to the temptation to begin there.

After expounding Putnam's deictic theory in his own way, the reviewer writes:

> But why should water beliefs, so understood, pose a problem for the functionalist? He will not be able to capture the content of such beliefs by alluding to their characteristic causal connections with experience alone. Rather, the important connections will be those between the current water thought, a past demonstrative thought about certain samples then in front of the subject, and the theoretical belief that such substances fall under some fundamental classification which may be revealed by investigating the causal properties of those substances. It is these connections which will ensure that the thought is indeed about water and they are part of that thought's causal role. (David Owens, *loc. cit.*)

I read that once and then, with the words "water beliefs" and "current water thoughts" and the ambiguity of the latter, something seemed to stir in memory. I read the passage again and I got the thing I was searching for. It was a footnote of J. L. Austin's: "There will not be books in the running brooks until the dawn of hydrosemantics."[7] It must be a weakness to be as easily amused as I am by such things. But now that (for "water" at least) we have got ourselves a hydrosemantics—now that the running brooks are in the books[8] (even if not yet vice versa)—I suggest that it is time to consider the question why in the 1950s and well into the 1960s it seemed so irresistibly comical to contemplate the very idea of hydrosemantics. What was so unthinkable in the idea of a hydrosemantics even for "hydor" ("water")? The answer to the question is that this was unthinkable to most philosophers of Austin's era because, even though everyone knew that Quine had exposed a grave circularity in all extant attempts to explain from a standing start the analytic/synthetic distinction, that did not seem to matter very much. It did not matter because analytic philosophers thought that, in practice, they could still reach principled consensus on what was analytic. So what good reason was there to question the separation of language from the world, to contemplate letting the running brooks into the books, or even to let water itself into the semantics of "water"? What reason was there to modify the idea that philosophy consisted in analysis? Analysis and

[7] J. L. Austin, 'Truth', *Proc. Ar. Soc.* Supp. vol. 24 (1950), 121.

[8] For a closer approximation to the hydrosemantics or natural meanings that the exiled Duke must really have had in mind at *As You Like It* II. i. 16, cf. Paul Valéry's celebration (however ill-calculated to evoke Austin's approval) of the Source Perrier.

the search for non-circular necessary and sufficient conditions could continue.[9]

By attacking the very idea of analyticity (a supposedly salutary exaggeration), the Quinean onslaught had overextended itself and miscarried. The point about Putnam's 'Is Semantics Possible?' is that it was part of a more modest, and far more effective, second onslaught on the idea of analyticity. This second attack is not targeted on the existence of definable single criterion concepts, *vixen* (= *female fox*), *oculist* (=*eye-doctor*). Nor is it an attack on the analyticity of "a vixen is a female fox", "an oculist is an eye-doctor", etc. That is scarcely worth attacking. What it is targeted upon is the attempt to *generalize* from those however undeniable small successes. "A theory which correctly describes the behaviour of perhaps three hundred words has been asserted to correctly describe the behaviour of the tens of thousands of general names."[10] In other words, what obstructed the discovery (or rediscovery, as I shall claim) of the necessity for a deictic view of natural kind names was the failure of our philosophical community at large to think through exactly and to the end the problems and limitations of the analytic/synthetic distinction. It is no accident that Putnam, who had spent as long as anyone in re-examining the distinction, should have been the rediscoverer (or co-rediscoverer with Kripke) of the deictic view of natural kind words. What I think *is* strange, if anything is, is that these matters have still not been thought right through[11] and have still not impinged as they should on the practice of philosophy. It is even stranger still, I think, that in the philosophy of science those who now concern themselves with the meaning of theoretical terms still show so little inclination to

[9] Indeed it still continues, to judge by the aims and ambitions that are prescribed by most of the participants in the group effort to solve for *x*—if necessary by brute force—in the equation: (knowledge) = (belief + x). In easel painting, pointillisme was a short-lived experiment. In philosophy, it bids fair to continue for ever.

[10] Kiefer and Munitz (eds.), *Language, Belief and Metaphysics*, 52.

[11] Not anticipating that Kripke would be charged with violating some supposed distinction between metaphysics and the philosophy of language (another case of the distinction Quine and Putnam subverted) and, falsely supposing that, after Kripke's excellent observations on the differences between the statuses of necessity, analyticity, and a priority, all these things would inevitably be thought through to the end, I rashly elected in *Sameness and Substance* to call certain truths and necessities that were neither analytically nor formally nor combinatorially guaranteed, but rested on what things (objects or kinds) both individuatively are and cannot help but be (that is on the individuation of objects under certain concepts of the sort Putnam had described), *conceptual truths*. I little thought that my "conceptual" would be read as an evasive synonym of "analytic", or that I would be seen as seeking to save some of the most implausible theses of linguistic philosophy. Why should a concept such as *man*, *horse*, *tree* be something that arises on the language side of a barrier that keeps the world from flowing into the word? It is expressly denied in *Sameness and Substance* that language can be protected by such an exclusion zone.

emancipate themselves from the purely model-theoretic approach or to explore the possibility of sense-giving relations between predicates and kinds that might mimic the direction of fit we now take for granted in our understanding of the setting up of the relation of designation. (See again paragraph 3, Section 2 above.)

I believe that the explanation of all this has something to do with the self-declared nature of empiricism and the unconsciousness of our commitment to it. But the point is hard to pin down, and it is now time for me to redeem my promise to show how the extension-involvingness of natural kind words will cohere with the theory of sense and reference.

6. On 24 May 1891 Frege wrote Husserl a letter about the sense and reference of predicates, and he included in it an instructive and remarkable diagram.

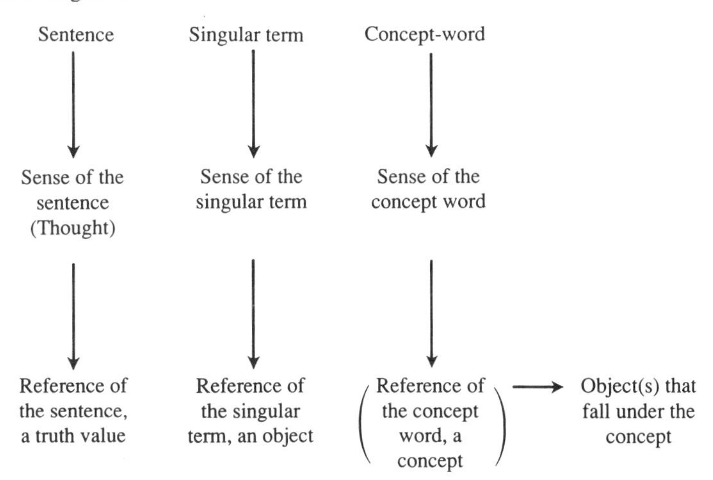

Let me first gloss this diagram in familiar ways. Then, after proposing a small repair to Fregean doctrine under the third column, let me try to show how neatly the deictic proposal can be assimilated within the scheme.

7. About column 1, let us simply remark that grasping the sense of a sentence is a matter of grasping under what conditions it attains to the True. In preference to saying that, in the strict sense of "designate" or "refer", the True is the reference or designation of a sentence, let us say that the True is the semantic value—Dummett's term—of the sentence. (Let us think of reference itself as a special case of semantic value.)

8. Now column 2. Here let us bear it in mind that, strictly and literally interpreted, the claim that "sense determines reference" is compatible with the denial of the *priority* of either.

Because the remark is rarely quoted, let me start by recalling to you Frege's remark, in *Ausführungen uber Sinn und Bedeutung*, about the fictional names "Nausicaa" and "moly". In so far as "Nausicaa" is allowed any sense, Frege says its literary signification depends upon its behaving *as if* (*als ob*) it designated some girl. (Similarly, the kind name "moly" gets what sense it has by behaving as if it designated a particular species of herb, namely the herb that in *Odyssey*, Book X, Hermes gives to Odysseus in preparation for his encounter with Circe. The name is still there, ready and waiting for any pharmaceutical product with similar or comparable apotropaic powers.) Now this is the case of fiction. But implicit in the claim about fiction is a simpler claim about the case of fact. The simpler claim is that in the non-fictional case a name has a sense by behaving as if it has—or simply by having—a reference. A name has its sense then by somehow presenting its object. To grasp the sense of a name is to know (in the manner correlative with the mode of presentation that corresponds to this particular somehow) which object the name is assigned to. To impart the sense is to show (in that manner) which object the name stands for.

To speak as Frege does of sense as "mode of presentation" suggests then nothing less than this: there is an object that the name presents and there is a way in which the name presents it, or a conception, as I shall say, of that object. A conception of an object is an account of how things are, a body of information (not necessarily correct in every particular) in which the object itself plays some distinct and distinguishable role.[12] Note, however, that such a conception or body of information will normally be open-ended, imperfect, and corrigible. Rarely, if ever, could it be condensed into a completed description of the object—but why should one ever have expected that there would be any description of x that is synonymous with a given name of x? What it is reasonable to expect is at most this: that the particular body of information about x to which the sense of some particular name of x is keyed may generate various different descriptions of x that might serve in a suitable context to identify x.

Properly possessed then, the conception of an object x that sustains

[12] See John McDowell, 'On the Sense and Reference of a Proper Name', *Mind*, 86 (1977), 59–85 (now Essay VII of this volume); David Wiggins, 'Frege's Problem of the Morning Star and the Evening Star', in M. Schirn (ed.), *Studies on Frege*, ii. *Logic and Philosophy of Language* (Stuttgart-Bad Canstatt: Frommann-Holzboog, 1976).

the sense of a name that is keyed to that conception is a way of thinking about x that fixes (with the help of the world, the world being what helps create the conception) which object the object in question is. Which object the object is is precisely what is mastered by him who comes to understand the name with the sense corresponding to that conception.

9. Now perhaps we can effect the transition to Frege's third column. Even as grasping the sense of a name or singular term and its contribution to truth conditions is grasping the particular conception of that object corresponding to the name's mode of presentation of what it stands for, and even as we give the term's sense by saying in a manner congruent with that conception what the term stands for, so grasping the sense of the predicate and its contribution to truth-conditions consists in grasping the predicate's mode of presentation of the concept it stands for, and we give the predicate's sense by saying what concept it stands for. And even as in the case of a singular term we show or exhibit this sense in preference to that sense by exploiting one mode of presentation rather than another to say which object this is—drawing upon one body of information in preference to another in filling out our identification of the object—so similarly, in the case of a predicate, we show or exhibit this sense in preference to that, e.g. the sense of "horse" rather than the sense of say "Equus caballus", by exploiting one mode of presentation rather than another to say which concept this is. We prefer (say) the body of information one might expound by saying "A horse is a certain animal with a flowing mane and tail; its voice is a neigh; and in the domestic state it is used as a beast of burden and draught, and for riding upon" over the body of information that identifies such creatures by classifying them as perissodactyl quadrupeds, locates their species among the genus *Equus* and the family *Equidae* and then dwells on other zoological features.

10. With this suggestion—entailing that there can be different accounts of one and the same kind of thing, namely *horse* or *Equus Caballus*—I can look forward to a possible conclusion about stereotypes, namely this: that Putnam's stereotypes approximate to those special conceptions that correspond to the various senses of the various expressions that stand for natural kinds; that stereotypes are particular special ways of thinking in identificatory fashion about such kinds and their specimens. That will indeed be my proposal. But there are confusions to be avoided here, and there is a departure from Putnam. We must hasten more slowly.

11. To avoid confusion, we must take care to present Frege's theory in column 3 as perfectly general with respect to all predicates and *then* say what is so special about natural kind words among predicates. It will also be necessary to indicate at some point what sort of thing Fregean concepts are. Finally, conceptions of a horse being in this picture conceptions *of* the concept horse (i.e. conceptions of what it is for something to be a horse), we must be as clear as possible about the difference between a conception, something belonging with sense, and a concept, which is something on the level of reference—lest we fall into the old error of confusing sense and reference.

So let us return to the analogy between names in general and predicates in general, that is the analogy between columns 2 and 3. If we take the analogy in the manner proposed, then it makes good sense of Frege's insistence, in the letter to Husserl and elsewhere, that the reference of a predicate cannot be any object or objects that it is true of. Just as singular terms without reference are unfitted to figure in the expression of a judgement possessed of a significance that enables us to move forward, as we must, to a truth value, just as a name capable of figuring in the expression of a judgement that can constitute knowledge *must* have a reference, so must any predicate that aspires to this status. But many predicates essential to the expression of good information do not have anything they are true of. Therefore their Fregean reference is not any object they are true of. Thus, as Frege says in the letter cited,

> With a concept word, it takes one more step to reach the object than it does with a proper name, and this last step may be missing—i.e. the concept may be empty—without the concept word's ceasing to be scientifically useful. I have drawn the last step, from concept to object, horizontally in order to indicate that it takes place on the same level, that objects and concepts have the same objectivity. In literary use it is sufficient if everything has sense; in scientific use there must also be *Bedeutungen*.

So on this account of things, the claim that a predicate has a concept as its reference does not bring out what is distinctive of natural kind terms or of Putnam's proposal concerning them. "Round square" has a reference, "blue" has a reference, "pencil" has a reference. You come to know this reference, which is nothing other than the concept, by coming to know what it would take (whether that be possible or impossible) for a thing to be a round square, blue, or a pencil. And now, *as one special case of this*, there are terms, such as "lemon" or "tiger", where to grasp what it would take for something to be a lemon or a tiger or whatever it is, you need exposure to the extension of the term. In this case—if the argument of 'Is Semantics Possible?' were correct—there is no grasping the reference or concept otherwise than through the extension. Here (at least) we cannot have reached the understanding that we do have by coming to grasp a strict lexical definition. For there is no strict lexical definition. So whereas the reality-involvingness of proper names amounts to their being reference-involving, the reality-involvingness of natural kind terms amounts to their being *extension-involving*. The schema itself of sense and reference neither demands that idea about extension nor excludes it, but it accommodates it.

12. On this account, we could say that he who understands "horse" and knows what Victor and Arkle both are, viz. horses, grasps a general rule for accepting or rejecting the sentence "*x* is a horse" according as it collects or fails to collect for arbitrary item *x* the verdict True. And, equally we could say that he who knows what Victor or Arkle are has got a grasp of the concept *horse*. These two accounts of the concept come to the same thing.[13]

At this point, I suppose I must digress for a moment to say one word about one well known and entirely general difficulty in Frege's scheme, a difficulty which has nothing to do with extension-involvingness, namely Frege's supposed need to deny that concept horse is a concept. If the view I take of these things is right, then the reference of "horse" is indeed something predicative but only in the following restricted sense. The term "horse" can be combined with the copula and article to give the predicative phrase "is a horse". The phrase "is a horse" is indeed, in Frege's terminology, an unsaturated expression. You can, if you wish, assign the phrase a semantic value. But to say that is not to say that the predicative phrase has a reference. The predicative phrase isn't what has the reference. "Horse" is what has the reference. In other words, the way back to the truth that the concept horse *is* a concept is to take the copula seriously and distinguish "horse" from "is a horse". If I say that there is something that Victor and Arkle both are, then the thing they both are is *horse*, not *is-a-horse*. What we quantify over are simply concepts. But a concept is the sort of thing which we can connect to a subject of discourse by linking a subject term and a predicate term by

[13] What then does it turn on whether concept C_1, is or is not the same as concept C_2? The question is difficult, but the difficulty is not one that we bring down on ourselves by exercising an escapable option to speak of concepts. Concepts are not philosophical artefacts. They are general things we are already committed to thinking about when we quantify (as we frequently do) over what predicates stand for. Once we see this, we shall not rush to offer any perfectly general answer, along the lines of Axiom V of Frege's *Grundgesetze*, or some modalization of this. (For continuants, we do not look for a general answer.) For some important contributions to the proper (that is the piecemeal) treatment of the problem, see Hilary Putnam, 'On Properties', in *Philosophical Papers*, i (Cambridge: Cambridge University Press, 1975).

the copula. The concept has a predicative role because it plays such a role in predication, but it is complete in itself, not unsaturated.[14]

There is much more to say on this subject, as there is on the analogy that Frege wanted to see between predicates and functors, but the time has come to sum up and move on to the next stage, which is to show how easily the deictic theory can be placed at column 3 within the framework of Fregean semantics.

13. In the case where we can only explain what is the reference of a predicate by something like the method that Putnam describes, because no strict analytical definition exists or could exist, the thing we who do know the sense have to recapitulate for someone else who does not know the sense of a given substantive is surely that which we have ourselves learned by commerce in the world at large with the objects that satisfy, or fall under, or exemplify the concept. In practice, and so in the theory of this practice, any specification of such a concept will have to depend directly or indirectly on exemplars. That is Putnam's point. But surely that point does not commit Putnam to think that the exemplars themselves can be given by bare unfocused presentation, unsupported by collateral explanation. (Cf. Section 4 above.) No presentation, one might say, without focus, and no focus without elucidation. When we fix the sense of an expression with a predicative role by giving its reference and give its reference by demonstrating or alluding to exemplars, what we need to impart to one who would learn the sense is both factual information *and* a practical capacity to recognize things of a certain kind, the information sustaining and regulating the recognition, and the recognition making possible the correction and amplification of the very information that first sustained the recognition. What we need to impart, we might say, is an identificatory or recognitional conception. It is this that corresponds to the sense of a natural kind term. What one must say then is that the sense of a natural kind term is correlative to a recognitional conception that is unspecifiable *except* as the conception of things like this, that and the other specimens exemplifying the concept that this conception is a conception of.

14. At this point it may be said that what I have just claimed isn't the same as what Putnam claimed. I shall attend to that in a moment. And there may be a more basic disquiet—disquiet at the fact that, if you

[14] For the semantics of the copula and other aspects of the difficulties that Frege encountered with the concept *horse*, see David Wiggins, 'The Sense and Reference of Predicates: A Plea for the Copula', in Crispin Wright (ed.), *Frege: Tradition and Influence* (Oxford: Blackwell, 1986), 126–43.

follow my suggestion about accommodating the doctrine of extension-involvingness, then you will find yourself saying: "that which supports the sense of the substantive 'horse' is a certain identificatory conception of the concept *horse*". Is this not an intolerable convergence in terminology? I shall attend to the first of these two worries in the course of allaying the second.

The convergence of "conception" and "concept" is ugly perhaps but it doesn't signify any confusion. It is clear what the terms "conception" and "concept" mean here, and it is clear what they have to do, namely, quite different work. But not only that. It is perfectly possible to replace the term "conception". There are at least three ways of doing this, each of them illuminating in its own way.

15. In the first place, we can replace the term "conception" by an expression of Evans's and we can say instead that there are two "ways of thinking about" the concept of horse or about horses. Either one can think about what Victor is as a familiar domestic animal (as a beast of burden, or mount) or one can think of what Victor is as a perissodactyl quadruped belonging to the genus *Equus* and the family *Equidae*. These are two different ways of thinking of one and the same kind of thing (concept).[15]

16. The second possibility is surely—is it not?—to replace the word conception by Putnam's word "stereotype". Or should this be doubted?

At the beginning of his article 'Is Semantics Possible?', Putnam introduced the deictic theory by saying that in order to impart the meaning of a thing-kind word one has to impart certain central facts, the stereotype and the extension. But what did Putnam mean by stereotype? And what did he think was the relation of stereotype and extension? When Putnam spoke of the stereotypes that support our normal understanding of the meaning of thing-kind words, he may sometimes have been thinking of something like the little engravings one finds in dictionaries like *Larousse*. (Indeed such engravings are full of theoretical interest in these connections. They are reminders of what deixis has to do in bringing language up to the world and the world up to language.) Normally, however, what Putnam means by a stereotype is a fund of ordinary information or a collection of idealized beliefs that one needs to grasp in order to get hold of the meaning of a thing-kind word. So a stereotype is rather like what you find—or one proper part of what you

[15] See Gareth Evans, *The Varieties of Reference*, ed. J. McDowell (Oxford: Clarendon Press, 1982), ch. 1.

find—in the text of a good dictionary or encyclopedia (which is not of course to say that it resembles an analytical definition).

What then must the relation be between the stereotype and the extension? Putnam himself was insistent that the core facts that determine the meaning of a thing-kind word were two perfectly separate kinds of fact, as if the stereotype could be explained or specified in a manner entirely independent of the question of its extension. The extension was the special province of experts and what belonged to it could be dark to possessors of the stereotype. But I find something strange in this separation. Of course the stereotype is different from the extension, but this is not to say that the first can be explained without any allusion to the second. To defend his own way of separating stereotype and extension Putnam himself claimed that he possessed the stereotype of elm and the stereotype of beech but didn't know the difference between an elm and a beech.[16] Reports have it that he has studiously defended this ignorance over twenty years in order to defend his opinion that one can understand a word by grasping the stereotype without having any grasp of what is the extension. Nevertheless, if Putnam is as vague as he says he is about the difference between an elm and beech, then I think one has a good right to doubt the degree of his comprehension of these words. If he entirely lacks the capacity to tell elms and beeches and to distinguish them, then there is something that he is missing. To insist that this thing he is missing is nothing *semantic* might become an upholder of the separation of language from the world or of the old understanding of the analytic–synthetic distinction. But these are things that Putnam's own work has done so much to discredit. Such insistence would be unbecoming issuing from him of all people. Would it not be better to say that the stereotype is the stereotype of this or that concept and that grasping the stereotype represents the beginning of an identificatory capacity, a capacity that the expert manifests more completely than the non-expert? In the case of the non-expert the capacity can be rudimentary, but surely it is identificatory. It is a capacity which could *advance* to the point where it became the capacity of an expert.

17. There is a third way to avoid saying that the conception which supports the sense of the word "horse" is a conception of the concept horse. This is to redraft one's description of the whole situation in terms inspired by Leibniz's theory of clear and distinct ideas. The thing we

[16] See 'The Meaning of "Meaning"', in *Philosophical Papers*, ii (Cambridge: Cambridge University Press, 1975).

most badly need here is what Leibniz called a clear but confused (i.e. non-distinct) idea.

In Leibniz's account of ordinary human knowledge, a clear idea of horse is not an image or a likeness of horse. It is that by the possession of which I recognize a horse when I encounter one. (What clarity in an idea contrasts with is not confusedness or non-distinctness, but obscurity.) A clear idea of horse is confused (or non-distinct) if, even though I can recognize a horse when I encounter one, I cannot enumerate one by one the marks which are sufficient to distinguish that kind of thing from another kind of thing. My understanding is simply practical and deictic. What I possess here I possess simply by having been brought into the presence of the thing. ("Being brought into the presence of a thing" translates Leibniz's own words.)[17] Our idea of horse will begin to become *distinct* as we learn to enumerate the marks that flow from the nature of a horse and that distinguish a horse from other creatures.

What Leibniz shows us how to describe here is nothing less than the process by which clear but indistinct knowledge of one and the same concept begins life anchored by a stereotype to examples that are grouped together by virtue of resemblances that are nomologically grounded. But then, as it grows, this knowledge gradually unfolds the concept in a succession of different and improving ideas (or conceptions—as we might have said). Rather than tell again the old Lockian story of nominal and real essence and face again the question what, if anything, makes them essences of some one and the same thing, we can now describe the process by which a clear indistinct idea becomes a clear distinct idea, then a clear adequate idea.[18]

This brings me to my last remark. When we reconstruct the first stages of the process, from the moment where there is ground to credit language-users with possession of a stereotype for horse, and ground to credit them with possession of a stereotype of the very same thing as we have a better and more informative stereotype for, and when we reflect that at that stage there could scarcely have *been* any experts, well we may think that strictly speaking, what Putnam should have stressed was not the necessity of experts but the necessity of the possibility of experts.

[17] See 'Meditationes de Cognitione, Veritate et Ideis' (Gerhardt IV), p. 422; *Discourse on Metaphysics* (ibid. Section 24–5); *New Essays*, 254–6.

[18] But only in a sense of "adequate" that must (I hold) be purged of certain Leibnizian preoccupations, e.g. the idea that, at the limit, as human knowledge approximates to God's knowledge, a posteriori knowledge will be able to be replaced by a priori demonstration.

More generally, and against the idea that extension-involvingness itself only reflects a stage in the development of scientific understanding, see *Sameness and Substance*, 210–13.

XII

THE CAUSAL THEORY OF NAMES

GARETH EVANS

I

1. In a paper which provides the starting point of this enquiry Saul Kripke opposes what he calls the Description Theory of Names and makes a counter-proposal of what I shall call the Causal Theory.[1] To be clear about what is at stake and what should be the outcome in the debate he initiated seems to me important for our understanding of talk and thought about the world in general as well as for our understanding of the functioning of proper names. I am anxious therefore that we identify the profound bases and likely generalizations of the opposing positions and do not content ourselves with counter-examples.

I should say that Kripke deliberately held back from presenting his ideas as a theory. I shall have to tighten them up, and I may suggest perhaps unintended directions of generalization; therefore his paper should be checked before the Causal Theory I consider is attributed to him.

There are two related but distinguishable questions concerning proper names. The first is about what the name denotes upon a particular occasion of its use when this is understood as being partly determinative of what the speaker strictly and literally said. I shall use the faintly barbarous coinage: *what the speaker denotes* (upon an occasion) for this notion. The second is about *what the name denotes*; we want to know what conditions have to be satisfied by an expression and an item for the first to be the, or a, name of the second. There is an entirely parallel pair of questions concerning general terms. In both cases it is ambiguity which prevents an easy answer of the first in terms of the second; to denote x it is not sufficient merely to utter something which is x's name.

Consequently there are two Description Theories, not distinguished by

From *Proceedings of the Aristotelian Society*, Supp. vol. 47 (1973), 187–208, by courtesy of the editor of the Aristotelian Society.

[1] S. A. Kripke, 'Naming and Necessity' (and Addenda), in D. Davidson and G. Harman (eds.), *Semantics of Natural Languages* (Dordrecht: D. Reidel, 1972), 253–355, 763–9.

Kripke.[2] The Description Theory of speaker's denotation holds that a name 'N.N.' denotes x upon a particular occasion of its use by a speaker S just in case x is uniquely that which satisfies all or most of the descriptions ϕ such that S would assent to 'N.N. is ϕ' (or '*That* N.N. is ϕ'). Crudely: the cluster of information S has associated with the name determines its denotation upon a particular occasion by *fit*. If the speaker has no individuating information he will denote nothing.

The Description Theory of what a name denotes holds that, associated with each name as used by a group of speakers who believe and intend that they are using the name with the same denotation, is a description or set of descriptions cullable from their beliefs which an item has to satisfy to be the bearer of the name. This description is used to explain the role of the name in existential, identity, and opaque contexts. The theory is by no means committed to the thesis that every user of the name must be in possession of the description; just as Kripke is not committed to holding that every user of the expression 'one metre' knows about the metre rod in Paris by saying that its reference is fixed by the description 'Length of stick S in Paris'. Indeed if the description is arrived at in the manner of Strawson[3]—averaging out the different beliefs of different speakers—it is most unlikely that the description will figure in every user's name-associated cluster.

The direct attack in Kripke's paper passes this latter theory by; most conspicuously the charge that the Description Theory ignores the social character of naming. I shall not discuss it explicitly either, though it will surface from time to time and the extent to which it is right should be clear by the end of the paper.

Kripke's direct attacks are unquestionably against the first Description Theory. He argues:

(a) An ordinary man in the street can denote the physicist Feynman by using the name 'Feynman' and say something true or false of him even though there is no description uniquely true of the physicist which he can fashion. (The conditions aren't necessary.)

(b) A person who associated with the name 'Gödel' merely the description 'prover of the incompleteness of Arithmetic' would none the less be denoting Gödel and saying something false of him in uttering 'Gödel proved the incompleteness of Arithmetic' even if an

[2] This can be seen in the way the list of theses defining the Description Theory alternate between those mentioning a speaker and those that don't, culminating in the uneasy idea of an idiolect of one. The Description Theorists of course do not themselves distinguish them clearly either, and many espouse both.

[3] P. F. Strawson, *Individuals* (London: Methuen, 1965), 191.

unknown Viennese by the name of Schmidt had in fact constructed the proof which Gödel had subsequently broadcast as his own. (If it is agreed that the speaker does not denote Schmidt the conditions aren't sufficient; if it is also agreed that he denotes Gödel, again they are not necessary.)

The strong thesis (that the Description Theorist's conditions are sufficient) is outrageous. What the speaker denotes in the sense we are concerned with is connected with saying in that strict sense which logicians so rightly prize, and the theory's deliverances of strict truth conditions are quite unacceptable. They would have the consequence, for example, that if I was previously innocent of knowledge or belief regarding Mr *Y*, and *X* is wrongly introduced to me as Mr *Y*, then I must speak the truth in uttering 'Mr *Y* is here' since *X* satisfies the overwhelming majority of descriptions I would associate with the name and *X* is here. I have grave doubts as to whether anyone has ever seriously held this thesis.

It is the weaker thesis—that some descriptive identification is necessary for a speaker to denote something—that it is important to understand. Strictly, Kripke's examples do not show it to be false since he nowhere provides a convincing reason for not taking into account speakers' possession of descriptions like 'man bearing such-and-such a name'; but I too think it is false. It can be seen as the fusion of two thoughts. First: that in order to be saying something by uttering an expression one must utter the sentence with certain intentions; this is felt to require, in the case of sentences containing names, that one be aiming at something with one's use of the name. Secondly—and this is where the underpinning from a certain Philosophy of Mind becomes apparent—to have an intention or belief concerning some item (which one is not in a position to demonstratively identify) one must be in possession of a description uniquely true of it. Both strands deserve at least momentary scrutiny.

We are prone to pass too quickly from the observation that neither parrots nor the wind *say* things to the conclusion that to say that *p* requires that one must intend to say that *p* and therefore, so to speak, be able to identify *p* independently of one's sentence. But the most we are entitled to conclude is that to say something one must intend to say something by uttering one's sentence (one normally will intend to say what it says). The application of the stricter requirement would lead us to relegate too much of our discourse to the status of mere mouthing. We constantly use general terms of whose satisfaction conditions we have but the dimmest idea. 'Microbiologist', 'chlorine' (the stuff in swimming pools), 'nicotine' (the stuff in cigarettes); these (and countless other words) we cannot define nor offer remarks which would distinguish their meaning from that of closely related words. It is wrong to say that we say nothing by uttering sentences containing these expressions, even if we recoil from the strong thesis, from saying that what we do say is determined by those hazy ideas and half-identifications we would offer if pressed.

The Philosophy of Mind is curiously popular but rarely made perfectly explicit.[4] It is held by anyone who holds that S believes that a is F if and only if

$$\exists\phi[(\text{S believes } \exists x(\phi x \ \& \ (\forall y)\ (\phi y \rightarrow x = y) \ \& \ \text{Fx})) \ \& \ \phi a \& \ (\forall y) \ (\phi y \rightarrow y = a)]$$

Obvious alterations would accommodate the other psychological attitudes. The range of the property quantifier must be restricted to exclude such properties as 'being identical with *a*'; otherwise the criterion is trivial.[5] The situation in which a thinking, planning, or wanting human has some item which is the object of his thought, plan, or desire is represented as a species of essentially the same situation as that which holds when there is no object and the thought, plan, or desire is, as we might say, purely general. There are thoughts, such as the thought that there are 11-fingered men, for whose expression general terms of the language suffice. The idea is that when the psychological state involves an object, a general term believed to be uniquely instantiated and in fact uniquely instantiated by the item which is the object of the state will figure in its specification. This idea may be coupled with a concession that there are certain privileged objects to which one may be more directly related; indeed such a concession appears to be needed if the theory is to be able to allow what appears an evident possibility: object-directed thoughts in a perfectly symmetrical or cyclical universe.

This idea about the nature of object-directed psychological attitudes obviously owes much to the feeling that there must be something we can say about what is believed or wanted even when there is no appropriate object actually to be found in the world. But it can also be seen as deriving support from a Principle of Charity: so attribute objects to beliefs that true belief is maximized. (I do not think this is an acceptable

[4] For example, see J. R. Searle, *Speech Acts* (Cambridge: Cambridge University Press, 1969), 87; E. Gellner, 'Ethics and Logic', *Proceedings of the Aristotelian Society*, 1954–5; B. Russell, *Problems of Philosophy* (Oxford: Oxford University Press, 1912), 29. E. Sosa criticizes it in 'Quantifiers, Belief and Sellars', in Davis, Hockney, and Wilson (eds.), *Philosophical Logic* (Dordrecht: D. Reidel, 1969), 69.

[5] I owe this observation to G. Harman.

principle; the acceptable principle enjoins minimizing the attribution of *inexplicable* error and therefore cannot be operated without a theory of the causation of belief for the creatures under investigation.)

We cannot deal comprehensively with this Philosophy of Mind here. My objections to it are essentially those of Wittgenstein. For an item to be the object of some psychological attitude of yours may be simply for you to be placed in a context which relates you to that thing. What makes it one rather than the other of a pair of identical twins that you are in love with? Certainly not some specification blueprinted in your mind; it may be no more than this: it was one of them and not the other that you have met. The theorist may gesture to the description 'the one I have met' but can give no explanation for the impossibility of its being outweighed by other descriptions which may have been acquired as a result of error and which may in fact happen to fit the other, unmet, twin. If God had looked into your mind, he would not have seen there with whom you were in love, and of whom you were thinking.

With that I propose to begin considering the Causal Theory.

2. The Causal Theory as stated by Kripke goes something like this. A speaker, using a name 'NN' on a particular occasion will denote some item x if there is a causal chain of *reference-preserving links* leading back from his use on that occasion ultimately to the item x itself being involved in a name-acquiring transaction such as an explicit dubbing or the more gradual process whereby nicknames stick. I mention the notion of a reference-preserving link to incorporate a condition that Kripke lays down; a speaker S's transmission of a name 'NN' to a speaker S' constitutes a reference-preserving link only if S intends to be using the name with the same denotation as he from whom he in his turn learned the name.

Let us begin by considering the theory in answer to our question about speaker's denotation (i.e., at the level of the individual speaker). In particular, let us consider the thesis that it is *sufficient* for someone to denote x on a particular occasion with the name that this use of the name on that occasion be a causal consequence of his exposure to other speakers using the expression to denote x.

An example which might favourably dispose one towards the theory is this. A group of people are having a conversation in a pub, about a certain Louis of whom S has never heard before. S becomes interested and asks: 'What did Louis do then?' There seems to be no question but that S denotes a particular man and asks about him. Or on some subsequent occasion S may use the name to offer some new thought to one of the participants: 'Louis was quite right to do that.' Again he clearly denotes whoever was the subject of conversation in the pub. This is difficult to reconcile with the Description Theory since the scraps of information which he picked up during the conversation might involve some distortion and fit someone else much better. Of course he has the description 'the man they were talking about' but the theory has no explanation for the impossibility of its being outweighed.

The Causal Theory can secure the right answer in such a case but I think deeper reflection will reveal that it too involves a refusal to recognize the insight about contextual determination I mentioned earlier. For the theory has the following consequence: that at any future time, no matter how remote or forgotten the conversation, no matter how alien the subject matter and confused the speaker, *S* will denote one particular Frenchman—perhaps Louis XIII—so long as there is a causal connexion between his use at that time and the long distant conversation.

It is important in testing your intuitions against the theory that you imagine the predicate changed—so that he says something like 'Louis was a basketball player' which was not heard in the conversation and which arises as the result of some confusion. This is to prevent the operation of what I call the 'mouthpiece syndrome' by which we attach sense and reference to a man's remarks only because we hear someone else speaking through him; as we might with a messenger, carrying a message about matters of which he was entirely ignorant.

Now there is no knock-down argument to show this consequence unacceptable; with pliant enough intuitions you can swallow anything in philosophy. But notice how little *point* there is in saying that he denotes one French King rather than any other, or any other person named by the name. There is now nothing that the speaker is prepared to say or do which relates him differentially to that one King. This is why it is so outrageous to say that he believes that Louis XIII is a basketball player. The notion of saying has simply been severed from all the connexions that made it of interest. Certainly we did not think we were letting ourselves in for this when we took the point about the conversation in the pub. What has gone wrong?[6]

The Causal Theory again ignores the importance of surrounding context, and regards the capacity to denote something as a magic trick which has somehow been passed on, and once passed on cannot be lost. We should rather say: in virtue of the context in which the man found

[6] Kripke expresses doubts about the sufficiency of the conditions for this sort of reason; see 'Naming and Necessity', 303.

himself the man's dispositions were bent towards one particular man—Louis XIII—whose states and doings alone he would count as serving to verify remarks made in that context using the name. And of course that context can persist, for the conversation can itself be adverted to subsequently. But it can also disappear so that the speaker is simply not sensitive to the outcome of any investigations regarding the truth of what he is said to have said. And at this point saying becomes detached, and uninteresting.

(It is worth observing how ambivalent Kripke is on the relation between denoting and believing; when the connexion favours him he uses it; we are reminded for example that the ordinary man has a false belief about Gödel and not a true belief about Schmidt. But it is obvious that the results of the 'who are they believing about?' criterion are bound to come dramatically apart from the results of the 'who is the original bearer of the name?' criterion, if for no other reason than that the former must be constructed to give results in cases where there is no name and where the latter cannot apply. When this happens we are sternly reminded that 'X refers' and 'X says' are being used in *technical* senses.[7] But there are limits. One could regard the aim of this essay to restore the connexion which must exist between strict truth conditions and the beliefs and interests of the users of the sentences if the technical notion of strict truth conditions is to be of interest to us.)

Reflection upon the conversation in the pub appeared to provide one reason for being favourably disposed towards the Causal Theory. There is another connected reason we ought to examine briefly. It might appear that the Causal Theory provides the basis for a general non-intentional answer to the Problem of Ambiguity. The problem is clear enough: What conditions have to be satisfied for a speaker to have said that *p* when he utters a sentence which may appropriately be used to say that *q* and that *r* and that *s* in addition? Two obvious alternative answers are

(a) the extent to which it is reasonable for his audience to conclude that he was saying that *p*
(b) his intending to say that *p*

and neither is without its difficulties. We can therefore imagine someone hoping for a natural extension of the Causal Theory to general terms which would enable him to explain for example how a child who did not have determinative intentions because of the technical nature of the

[7] Ibid. 348 n.

subject-matter may still say something determinate using a sentence which is in fact ambiguous.

I touch upon this to ensure that we are keeping the range of relevant considerations to be brought to bear upon the debate as wide as it must be. But I think little general advantage can accrue to the Causal Theory from thus broadening the considerations. The reason is that it simply fails to have the generality of the other two theories; it has no obvious application, for example, to syntactic ambiguity or to ambiguity produced by attempts to refer with non-unique descriptions, or pronouns. It seems inconceivable that the general theory of disambiguation required for such cases would be inadequate to deal with the phenomenon of shared names and would require *ad hoc* supplementation from the Causal Theory.

I want to stress how, precisely because the Causal Theory ignores the way context can be determinative of what gets *said*, it has quite unacceptable consequences. Suppose for example on a TV quiz programme I am asked to name a capital city and I say 'Kingston is the capital of Jamaica'; I should want to say that I had said something strictly and literally true even though it turns out that the man from whom I had picked up this scrap of information was actually referring to Kingston-upon-Thames and making a racist observation.

It may begin to appear that what gets said is going to be determined by what name is used, what items bear the name, and general principles of contextual disambiguation. The causal origin of the speaker's familiarity with the name, save in certain specialized 'mouthpiece cases', does not seem to have a critical role to play.

This impression may be strengthened by the observation that a causal connexion between my use of the name and use by others (whether or not leading back ultimately to the item itself) is simply not necessary for me to use the name to say something. Amongst the Wagera Indians, for example, 'newly born children receive the names of deceased members of their family according to fixed rules . . . the first born takes on the name of the paternal grandfather, the second that of the father's eldest brother, the third that of the maternal grandfather.'[8] In these and other situations (names for streets in US cities, etc.) a knowledgeable speaker may excogitate a name and use it to denote some item which bears it without any causal connexion whatever with the use by others of that name.

These points might be conceded by Kripke while maintaining the

[8] E. Delhaise, 'Les Wagera', *Monogr. Ethnogr.* (1909).

general position that the denotation of a name in a community is still to be found by tracing a causal chain of reference preserving links back to some item. It is to this theory that I now turn.

3. Suppose a parallel theory were offered to explain the sense of general terms (not just terms for natural kinds). One would reply as follows: 'There aren't two fundamentally different mechanisms involved in a word's having a meaning: one bringing it about that a word acquires a meaning, and the other—a causal mechanism—which operates to ensure that its meaning is preserved. The former processes are operative all the time; whatever explains how a word gets its meaning also explains how it preserves it, if preserved it is. Indeed such a theory could not account for the phenomenon of a word's changing its meaning. It is perfectly possible for this to happen without anyone's intending to initiate a new practice with the word; the causal chain would then lead back too far.'

Change of meaning would be decisive against such a theory of the meaning of general terms. Change of denotation is similarly decisive against the Causal Theory of Names. Not only are changes of denotation imaginable, but it appears that they actually occur. We learn from Isaac Taylor's book: *Names and their History*, 1898: "In the case of 'Madagascar' a hearsay report of Malay or Arab sailors misunderstood by Marco Polo . . . has had the effect of transferring a corrupt form of the name of a portion of the African mainland to the great African Island." A simple imaginary case would be this: Two babies are born, and their mothers bestow names upon them. A nurse inadvertently switches them and the error is never discovered. It will henceforth undeniably be the case that the man universally known as 'Jack' is so called because a woman dubbed some other baby with the name.

It is clear that the Causal Theory unamended is not adequate. It looks as though, once again, the intentions of the speakers to use the name to refer to something must be allowed to count in determination of what it denotes.

But it is not enough to say that and leave matters there. We must at least sketch a theory which will enable 'Madagascar' to be the name of the island yet which will not have the consequence that 'Gödel' would become a name of Schmidt in the situation envisaged by Kripke nor 'Goliath' a name of the Philistine killed by David. (Biblical scholars now suggest that David did not kill Goliath, and that the attribution of the slaying to Elhannan the Bethlehemite in 2 Samuel 21: 19 is correct. David is thought to have killed a Philistine but not Goliath.[9]) For

[9] Henry Wheeler Robinson, *History of Israel*, rev. edn. (Naperville, Ill.: Allenson, 1964), 187.

although this has never been explicitly argued I would agree that even if the 'information' connected with the name in possession of an entire community was merely that 'Goliath was the Philistine David slew' this would still not mean that 'Goliath' referred in that community to that man, and therefore that the sentence expressed a truth. And if we simultaneously thought that the name *would* denote the Philistine slain by Elhannan then both the necessity and sufficiency of the conditions suggested by the Description Theory of the denotation of a name are rejected. This is the case Kripke should have argued but didn't.

4. Before going on to sketch such a theory in the second part of this essay let me survey the position arrived at and use it to make a summary statement of the position I wish to adopt.

We can see the undifferentiated Description Theory as the expression of two thoughts.

(a) the denotation of a name is determined by what speakers intend to refer to by using the name.

(b) the object a speaker intends to refer to by his use of a name is that which satisfies or fits the majority of descriptions which make up the cluster of information which the speaker has associated with the name.

We have seen great difficulties with (a) when this is interpreted as a thesis at the micro level. But consideration of the phenomenon of a name's getting a denotation, or changing it, suggests that there being a community of speakers using the name with such and such as the intended referent is likely to be a crucial constituent in these processes. With names as with other expressions in the language, what they signify depends upon what we use them to signify; a truth whose recognition is compatible with denying the collapse of saying into meaning at the level of the individual speaker.

It is in (b) that the real weakness lies: the bad old Philosophy of Mind which we momentarily uncovered. Not so much in the idea that the intended referent is determined in a more or less complicated way by the associated information, but the specific form the determination was supposed to take: *fit*. There is something absurd in supposing that the intended referent of some perfectly ordinary use of a name by a speaker could be some item utterly isolated (causally) from the user's community and culture simply in virtue of the fact that it fits better than anything else the cluster of descriptions he associates with the name. I would agree with Kripke in thinking that the absurdity resides in the absence of any causal relation between the item concerned and the speaker. But it seems to me that he has mislocated the causal relation; the important causal relation lies between that item's states and doings and the speaker's body of information—not between the item's being dubbed with a name and the speaker's contemporary use of it.

Philosophers have come increasingly to realize that major concepts in epistemology and the philosophy of mind have causality embedded within them. Seeing and knowing are both good examples.

The absurdity in supposing that the denotation of our contemporary use of the name 'Aristotle' could be some unknown (n.b.) item whose doings are causally isolated from our body of information is strictly parallel to the absurdity in supposing that one might be seeing something one has no causal contact with solely upon the ground that there is a splendid match between object and visual impression.

There probably is some *degree of fit* requirement in the case of seeing which means that after some amount of distortion or fancy we can no longer maintain that the causally operative item was still being seen. And I think it is likely that there is a parallel requirement for referring. We learn for example from E. K. Chambers' *Arthur of Britain* that Arthur had a son Anir "whom legend has perhaps confused with his burial place". If Kripke's notion of reference fixing is such that those who said Anir was a burial place of Arthur might be denoting a person it seems that it has little to commend it, and is certainly not justified by the criticisms he makes against the Description Theory. But the existence or nature of this 'degree of fit' requirement will not be something I shall be concerned with here.

We must allow then that the denotation of a name in the community will depend in a complicated way upon what those who use the term intend to refer to, but we will so understand 'intended referent' that typically a *necessary* (but not sufficient) condition for x's being the intended referent of S's use of a name is that x should be the source of causal origin of the body of information that S has associated with the name.

II

5. The aim I have set myself, then, is modest; it is not to present a complete theory of the denotation of names. Without presenting a general theory to solve the problem of ambiguity I cannot present a theory of speaker's denotation, although I will make remarks which prejudice that issue. I propose merely to sketch an account of what makes an expression into a name for something that will allow names to change their denotations.

The enterprise is more modest yet for I propose to help myself to an undefined notion of speaker's reference by borrowing from the theory of communication. But a word of explanation.

A speaker may have succeeded in *getting it across* or in *communicating* that p even though he uses a sentence which may not appropriately be used to say that p. Presumably this success consists in his audience's having formed a belief about him. This need not be the belief that the speaker intended to say in the strict sense that p, since the speaker may succeed in getting something across despite using a sentence which he is known to know cannot appropriately be used to say that p. The speaker will have referred to a, in the sense I am helping myself to, only if he has succeeded in getting it across that Fa (for some substitution F). Further stringent conditions are required. Clearly this notion is quite different from the notion of denotation which I have been using, tied as denotation is to saying in the strict sense. One may refer to x by using a description that x does not satisfy; one may not thus denote x.

Now a speaker may know or believe that there is such-and-such an item in the world and intend to refer to it. And this is where the suggestion made earlier must be brought to bear, for *that* item is not (in general) the satisfier of the body of information the possession by the speaker of which makes it true that he knows of the existence of the item; it is rather that item which is causally responsible for the speaker's possession of that body of information, or dominantly responsible if there is more than one. (The point is of course not specific to this intention, or to intention as opposed to other psychological attitudes.) Let us then, very briefly, explore these two ideas: source and dominance.

Usually our knowledge or belief about particular items is derived from information-gathering transactions, involving a causal interaction with some item or other, conducted ourselves or is derived, maybe through a long chain, from the transactions of others. Perception of the item is the main but by no means the only way an item can impress itself on us; for example, a man can be the source of things we discover by rifling through his suitcase or by reading his works.

A causal relation is of course not sufficient; but we may borrow from the theory of knowledge and say something like this. X is the source of the belief S expresses by uttering 'Fa' if there was an episode which caused S's belief in which X and S were causally related in a type of situation apt for producing knowledge that something F-s ($\exists x(Fx)$)—a type of situation in which the belief that something F-s would be caused by something's F-ing. That it is a way of producing knowledge does not mean that it cannot go wrong; that is why X, by smoking French cigarettes, can be the source of the belief S expresses by '*a* smokes Greek cigarettes'.

Of course some of our information about the world is not so based; we may deduce that there is a tallest man in the world and deduce that he is over 6 feet tall. No man is the source of this information; a name introduced in relation to it might function very much as the unamended Description Theory suggested.

Legend and fancy can create new characters, or add bodies of sourceless material to other dossiers; restrictions on the causal relation would prevent the inventors of the legends turning out to be the sources of the beliefs their legends gave rise to. Someone other than the ϕ can be the source of the belief S expresses by '*a* is the ϕ'; Kripke's Gödel, by claiming the proof, was the source of the belief people manifested by saying 'Gödel proved the incompleteness of Arithmetic', not Schmidt.

Misidentification can bring it about that the item which is the source of the information is different from the item about which the information is believed. I may form the belief about the wife of some colleague that she has nice legs upon the basis of seeing someone else—but the girl I saw is the source.

Consequently a cluster or dossier of information can be dominantly *of*[10] an item though it contains elements whose source is different. And we surely want to allow that persistent misidentification can bring it about that a cluster is dominantly of some item other than that it was dominantly of originally.

Suppose I get to know a man slightly. Suppose then a suitably primed identical twin takes over his position, and I get to know him fairly well, not noticing the switch. Immediately after the switch my dossier will still be dominantly of the original man, and I falsely believe, as I would acknowledge if it was pointed out, that *he* is in the room. Then I would pass through a period in which neither was dominant; I had not misidentified one as the other, an asymmetrical relation, but rather confused them. Finally the twin could take over the dominant position; I would not have false beliefs about who is in the room, but false beliefs about e.g., when I first met the man in the room. These differences seem to reside entirely in the differences in the believer's reactions to the various discoveries, and dominance is meant to capture those differences.

[10] The term is D. Kaplan's; see 'Quantifying In', in *Words and Objections*, ed. D. Davidson and J. Hintikka (New York: Humanities Press, 1969). I think there are clear similarities between my notion of a dominant source and notions he is there sketching. However I want nothing to do with vividness. I borrow the term "dossier" from H. P. Grice's paper 'Vacuous Names' in the same volume.

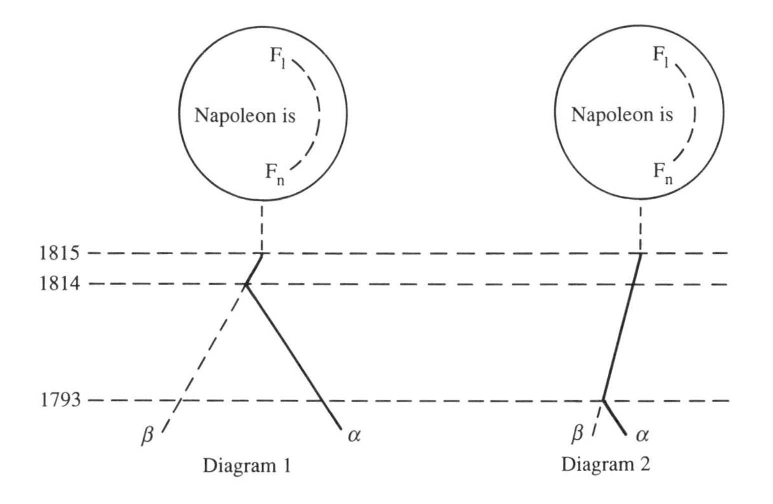

Dominance is not simply a function of *amount* of information (if that is even intelligible). In the case of persons, for example, each man's life presents a skeleton and the dominant source may be the man who contributed to covering most of it rather than the man who contributed most of the covering. Detail in a particular area can be outweighed by spread. Also the believer's reasons for being interested in the item at all will weigh.

Consider another example. If it turns out that an impersonator had taken over Napoleon's role from 1814 onwards (post-Elba) the cluster of the typical historian would still be dominantly of the man responsible for the earlier exploits (α in diagram 1) and we would say that they had false beliefs about who fought at Waterloo. If however the switch had occurred much earlier, it being an unknown Army Officer being impersonated, then their information would be dominantly of the later man (β in diagram 2). They did not have false beliefs about who was the general at Waterloo, but rather false beliefs about that general's early career.

I think we can say that *in general* a speaker intends to refer to the item that is the dominant source of his associated body of information. It is important to see that this will not change from occasion to occasion depending upon subject-matter. Some have proposed[11] that if in case

[11] K. S. Donnellan, 'Proper Names and Identifying Descriptions', in Davidson and Harman (eds.), *Semantics of Natural Language*, 371.

1 the historian says 'Napoleon fought skilfully at Waterloo' it is the impostor β who is the intended referent, while if he had said in the next breath '. . . unlike his performance in the Senate' it would be α. This seems a mistake; not only was what the man said false, what he intended to say was false too, as he would be the first to agree; it wasn't Napoleon who fought skilfully at Waterloo.

With this background then we may offer the following tentative definition:

'NN' is a name of x if there is a community C

1. in which it is common knowledge that members of C have in their repertoire the procedure of using 'NN' to refer to x (with the intention of referring to x)
2. the success in reference in any particular case being intended to rely on common knowledge between speaker and hearer that 'NN' has been used to refer to x by members of C and not upon common knowledge of the satisfaction by x of some predicate embedded in 'NN'.[12]

(In order to keep the definition simple no attempt is made to cover the sense in which an unused institutionally-approved name is a name.)

This distinction (between use-because-(we know)-we-use-it and use upon other bases) is just what is needed to distinguish dead from live metaphors; it seems to me the only basis on which to distinguish the referential functioning of names, which may grammatically be descriptions.[13]

The definition does not have the consequence that the description 'the man we call "NN"' is a name, for *its* success as a referential device does not rely upon common knowledge that *it* is or has been used to refer to x.

Intentions alone don't bring it about that a name gets a denotation; without the intentions being manifest there cannot be the common knowledge required for the practice.

Our conditions are more stringent than Kripke's since for him an expression becomes a name just so long as someone has dubbed something with it and thereby caused it to be in common usage. This seems

[12] For the notion of "common knowledge", see D. K. Lewis, *Convention* (Cambridge, Mass.: Harvard University Press, 1969), and the slightly different notion in S. Schiffer, *Meaning* (Oxford: Clarendon Press, 1972). For the notion of "a procedure in the repertoire", see H. P. Grice, 'Utterer's Meaning, Sentence Meaning, Word Meaning', *Foundations of Language* (1968). Clearly the whole enterprise owes much to Grice but no commitment is here made to any specific version of the theory of communication.

[13] And if Schiffer is right much more as well—see *Meaning*, ch. 5.

little short of magical. Suppose one of a group of villagers dubbed a little girl on holiday in the vicinity 'Goldilocks' and the name caught on. However suppose that there were two identical twins the villagers totally fail to distinguish. I should deny that 'Goldilocks' is the name of either—even if by some miracle each villager used the name consistently but in no sense did they fall into two coherent sub-communities. (The name might denote the girl first dubbed if for some peculiar reason the villagers were deferential to the introducer of the name—of this more below.)

Consider the following case. An urn is discovered in the Dead Sea containing documents on which are found fascinating mathematical proofs. Inscribed at the bottom is the name 'Ibn Khan' which is quite naturally taken to be the name of the constructor of the proofs. Consequently it passes into common usage amongst mathematicians concerned with that branch of mathematics. 'Khan conjectured here that . . .' and the like. However suppose the name was the name of the scribe who had transcribed the proofs much later; a small '*id scripsit*' had been obliterated.

Here is a perfect case where there is a coherent community using the name with the mathematician as the intended referent and a consequence of the definition would be that 'Ibn Khan' would be one of his names. Also, 'Malachai' would have been the name of the author of the Biblical work of the same name despite that its use was based upon a misapprehension ('Malachai' means my messenger).[14]

Speakers within such traditions use names under the misapprehension that their use is in conformity with the use of other speakers referring to the relevant item. The names would probably be withdrawn when that misapprehension is revealed, or start a rather different life as "our" names for the items (cf. 'Deutero Isaiah' etc.). One might be impressed by this, and regard it as a reason for denying that those within these traditions spoke the literal truth in using the names. It is very easy to add a codicil to the definition which would have this effect.

Actually it is not a very good reason for denying that speakers within such traditions are speaking the literal truth.[15] But I do not want to insist upon any decision upon this point. This is because one can be concessive and allow the definition to be amended without giving up anything of importance. First: the definition with its codicil will still allow many names to change their denotation. Secondly: it obviously fails to follow

[14] See Otto Eissfeldt, *The Old Testament* (New York: Harper and Row, 1965), 441.

[15] John McDowell has persuaded me of this, as of much else. He detests my conclusions.

from the fact that, in our example, the community of mathematicians were not denoting the mathematician that they were denoting the scribe and were engaged in strictly speaking massive falsehood of him.

Let me elaborate the first of these points.

There is a fairly standard way in which people get their names. If we use a name of a man we expect that it originated in the standard manner and this expectation may condition our use of it. But consider names for people which are obviously nicknames, or names for places or pieces of music. Since there is no standard way in which these names are bestowed subsequent users will not in general use the name under any view as to its origin, and therefore when there is a divergence between the item involved in the name's origin and the speakers' intended referent there will be no *mis*apprehension, no latent motive for withdrawing the name, and thus no bar to the name's acquiring a new denotation even by the amended definition. So long as they have no reason to believe that the name has dragged any information with it, speakers will treat the revelation that the name had once been used to refer to something different with the same sort of indifference as that with which they greet the information that 'meat' once meant groceries in general.

We can easily tell the story in case 2 of our Napoleon diagram so that α was the original bearer of the name 'Napoleon' and it was transferred to the counterfeit because of the similarity of their appearances and therefore without the intention on anyone's part to initiate a new practice. Though this is not such a clear case I should probably say that historians have used the name 'Napoleon' to refer to β. They might perhaps abandon it, but that of course fails to show that they were all along denoting α. Nor does the fact that someone in the know might come along and say 'Napoleon was a fish salesman and was never at Waterloo' show anything. The relevant question is: 'Does this contradict the assertion that was made when the historians said "Napoleon was at Waterloo"?' To give an affirmative answer to this question requires the prior determination that they have all along been denoting α.

We need one further and major complication. Although standardly we use expressions with the intention of conforming with the general use made of them by the community, sometimes we use them with the *overriding* intention to conform to the use made of them by some other person or persons. In that case I shall say that we use the expression *deferentially* (with respect to that other person or group of persons). This is true of some general terms too: "viol", "minuet" would be examples.

I should say, for example, that the man in the conversation in the pub used 'Louis' deferentially. This is not just a matter of his ignorance; he could, indeed, have an opinion as to who this Louis is (the man he met earlier perhaps) but still use the expression deferentially. There is an important gap between

intending to refer to the ϕ and believing that a = the ϕ intending to refer to a

for even when he has an opinion as to whom they are talking about I should say that it was the man they were talking about, and not the man he met earlier, that he intended to refer to.

Archaeologists might find a tomb in the desert and claim falsely that it is the burial place of some little known character in the Bible. They could discover a great deal about the man in the tomb so that he and not the character in the Bible was the dominant source of their information. But, given the nature and point of their enterprise, the archaeologists are using the name deferentially to the authors of the Bible. I should say, then, that they denote that man, and say false things about him. Notice that in such a case there is some point to this characterization.

The case is in fact no different with any situation in which a name is used with the over-riding intention of referring to something satisfying such and such a description. Kripke gives the example of 'Jack the Ripper'. Again, after the arrest of a man a not in fact responsible for the crimes, a can be the dominant source of speakers' information but the intended referent could well be the murderer and not a. Again this will be productive of a whole lot of falsehood.

We do not use all names deferentially, least of all deferentially to the person from whom we picked them up. For example the mathematicians did not use the name 'Ibn Khan' with the *over-riding* intention of referring to whoever bore that name or was referred to by some other person or community.

We must thus be careful to distinguish two reasons for something that would count as "withdrawing sentences containing the name"

(a) the item's not bearing the name 'NN' ('Ibn Khan', 'Malachai')
(b) the item's not being NN (the biblical archaeologists).

I shall end with an example that enables me to draw these threads together and summarize where my position differs from the Causal Theory.

A youth *A* leaves a small village in the Scottish highlands to seek his fortune having acquired the nickname 'Turnip' (the reason for choosing a nickname is I hope clear). Fifty or so years later a man *B* comes to the village and lives as a hermit over the hill. The three or four villagers surviving from the time of the youth's departure believe falsely that this is the long departed villager returned. Consequently they use the name 'Turnip' among themselves and it gets into wider circulation among the younger villagers who have no idea how it originated. I am assuming that the older villagers, if the facts were pointed out, would say: 'It isn't Turnip after all' rather than 'It appears after all that Turnip did not come from this village'. In that case I should say that they use the name to refer to *A*, and in fact, denoting him, say false things about him (even by uttering 'Here is Turnip coming to get his coffee again').

But they may die off, leaving a homogeneous community using the name to refer to the man over the hill. I should say the way is clear to its becoming his name. The story is not much affected if the older villagers pass on some information whose source is *A* by saying such things as 'Turnip was quite a one for the girls', for the younger villagers' clusters would still be dominantly of the man over the hill. But it is an important feature of my account that the information that the older villagers gave the younger villagers could be so rich, coherent, and important to them that *A* could be the dominant source of their information, so that they too would acknowledge 'That man over the hill isn't Turnip after all'.

A final possibility would be if they used the name deferentially towards the older villagers, for some reason, with the consequence that no matter who was dominant they denote whoever the elders denote.

6. *Conclusion*. Espousers of both theories could reasonably claim to be vindicated by the position we have arrived at. We have secured for the Description Theorist much that he wanted. We have seen for at least the most fundamental case of the use of names (non-deferentially used names) the idea that their denotation is fixed in a more or less complicated way by the associated bodies of information that one could cull from the users of the name turns out not to be so wide of the mark. But of course that the fix is by causal origin and not by fit crucially affects the impact this idea has upon the statement of the truth conditions of existential or opaque sentences containing names. The theorist can also point to the idea of dominance as securing what he was trying, admittedly crudely, to secure with his talk of the 'majority of' the descriptions, and to the "degree of fit requirement" as blocking consequences he found objectionable.

The Causal Theorist can also look with satisfaction upon the result, incorporating as it does his insight about the importance of causality into a central position. Further, the logical doctrines he was concerned to establish, for example the non-contingency of identity statements made with the use of names, are not controverted. Information is individuated by source; if *a* is the source of a body of information nothing else could have been. Consequently nothing else could have been *that a*.

The only theorists who gain no comfort are those who, ignoring Kripke's explicit remarks to the contrary,[16] supposed that the Causal Theory could provide them with a totally *non-intentional* answer to the problem posed by names. But I am not distressed by their distress.

Our ideas also point forwards; for it seems that they, or some close relative, must be used in explaining the functioning of at least some demonstratives. Such an expression as 'That mountaineer' in 'That mountaineer is coming to town tonight' may advert to a body of information presumed in common possession, perhaps through the newspapers, which fixes its denotation. No one can be *that* mountaineer unless he is the source of that information no matter how perfectly he fits it, and of course someone can be that mountaineer and fail to fit quite a bit of it. It is in such generality that defence of our ideas must lie.

But with these hints I must leave the subject.

[16] 'Naming and Necessity', 302.

XIII

FREGE'S DISTINCTION BETWEEN SENSE AND REFERENCE

MICHAEL DUMMETT

Frege was the founder both of modern logic and of modern philosophy of language. The use of the latter phrase, in connection with him, has an odd ring, since he frequently expressed a contempt for language; he did so because, on his pen, 'language' meant 'natural language', and he believed that natural languages are very faulty instruments for the expression of thought. Not only are surface appearances, in sentences of natural language, grossly misleading, but, according to Frege, natural languages are incoherent in the sense that no complete systematic account of the use of the sentences of such a language could be framed. In our everyday speech, we are not merely playing a game whose rules we have not fully formulated, but one for which no consistent set of rules could be drawn up. Such an observation differs from what other philosophers have said about natural language only in its strength: where Frege diverged from his predecessors was in the methodological remedy he adopted. Others have thought that the philosopher's task is to divest thought of its linguistic clothing, to penetrate all forms of mere expression to the pure thought which lies beneath: Frege was the first to attach due weight to the fact that we cannot have a thought which we do not express, to ourselves if not to others. Any attempt to scrutinize our thoughts, taken apart from their expression, will therefore end in confusing the inner experience of thinking, or the merely contingent mental accompaniments of thinking, with the thoughts themselves. Thought differs from other things also said to be objects of the mind, for instance pains or mental images, in not being essentially private. I can tell you what my pain is like, or what I am visualizing, but I cannot transfer to you my pain or my mental image. It is of the essence of thought, however, that it is transferable, that I can convey to you exactly what I am thinking: as Wittgenstein said, in a passage critical of this conception of Frege's, you as it were take the thought into your mind; I do more

Originally published in Spanish as 'Frege' in *Teorema*, 5 (1975); published in this form in *Truth and Other Enigmas* (London: Duckworth, 1978).

than tell you what my thought is like—I communicate to you that very thought. Hence any attempt to investigate thoughts which culminates in a study of what is in essence private, that is, of inner mental experience, must have missed its mark.

The remedy which Frege adopted for the misleading and unsatisfactory character of our everyday languages was, therefore, to replace them, for purposes of philosophical study, by linguistic forms lacking these defects. Sentences of such an ideally constructed language would express thoughts in virtue of the principles governing the use of their constituent words, principles capable of systematic formulation. Because thought is communicable without residue, and because what is communicated depends only upon a common apprehension of the principles governing the language, such principles must relate only to the actual employment of sentences of the language, that is, to what in the use of the language is open to observation unaided by any supposed contact between mind and mind other than via the medium of language. It follows that, in order to study thought, our task is not to describe something mental in the sense of something lying outside the boundaries of the physical world; it is, rather, first to devise improved means for expressing thought, and, secondly, to formulate accurately and explicitly the principles governing the employment of such forms of expression—principles which we ordinarily leave merely implicit, and, in the case of natural language, could not be coherently formulated at all.

Frege's initial insight was that sentences play a primary role in the theory of meaning. A sentence is the smallest linguistic complex which one can use to *say* anything: hence the meaning of a word is to be given in terms of the contribution it makes to determining what may be said by means of a sentence containing it.

His second insight is that the notion of *truth* plays a crucial role in the account we have to give of the meaning of a sentence, and hence, by the first principle, of the meaning of any expression. In order to understand an assertoric utterance, we have both to know what it is to assert something, and what is the content of this particular assertion, that is, what condition must be fulfilled for what is said to be true. Likewise, to understand an interrogative utterance, we have both to know what it is to ask a question, and what is the content of this particular question, that is (for a question requiring the answer 'Yes' or 'No'), what is the condition under which it should receive the answer 'Yes'. We can assimilate the two by extending the notion of truth to cover interrogative sentences also, that is, by regarding them as true when the correct answer is 'Yes': if we do this, then, in both cases, the speaker utters a sentence which is true just in case a certain condition obtains; the difference consists wholly in what more the speaker is doing, namely, in the former case, asserting that the condition is fulfilled, and, in the latter, asking whether it is fulfilled. Frege refers to this difference as a difference in the *force* attached to the sentence in the two cases (assertoric or interrogative). Often some linguistic feature of the sentence will serve to indicate what force is attached to it, and, if we fail to understand the convention governing this indication of force, we shall not grasp the meaning of the sentence. Force is thus one aspect of meaning, to be distinguished from that ingredient of the meanings of the words which goes to determine the condition under which the sentence is true. This latter ingredient Frege calls the *sense*, and it is this about which I am going to speak.

The notion of truth is also, plainly, central for logic in the traditional sense of the science of inference. Logic began with Aristotle's discovery that the validity of an argument could be characterized by its being an instance of a valid argument-schema, where an argument-schema is like an argument save for containing schematic letters at certain places instead of actual words or expressions, and is valid if every instance with true premisses has a true conclusion, an instance being obtained by replacing each schematic letter with an actual expression of the appropriate logical category. This presemantic notion of an interpretation of a schema by *replacement* was the only one that logic had to operate with until Frege. Frege supplied us for the first time with a semantics, that is to say, an analysis of the way in which a sentence is determined as true or otherwise in accordance with its composition out of its constituent words. To arrive at a semantics, we need to carry out a preliminary task—to give a suitable account of the syntactical structure of our sentences, of how they are compounded out of words, Frege's dual achievement, which makes him the founder of modern logic, lay in his first giving such a syntactic analysis, and then laying down, in terms of it, a semantics. To speak more accurately, he did not directly give any syntactic analysis of natural language, but, instead, invented a formalized language whose sentences have a precise syntactic structure, and one which is open to view, since their surface appearance reveals it; and, having done so, provided a semantics for this language. It is rather generally supposed that we shall arrive at a satisfactory syntactic analysis of natural language only by exhibiting its sentences as having an underlying (or deep) structure analogous to that of sentences of Frege's formalized language, which has, of course, become the standard type of quantificational language employed by logicians. Frege himself made no such claim, supposing that natural language resists any complete coherent systematization; but it is on some version of such a claim, for at least a large fragment of natural language, that Frege's title to be the founder, not only of modern logic, but also of modern philosophy of language, rests.

At least this much is not seriously challenged by anyone: that a language having the sort of syntax that Frege devised for his formalized language is adequate for the formulation of any theory of mathematics or natural science; disputes arise only over forms of sentence-construction important for other purposes. Frege's syntax is based on the fundamental idea that the construction of a sentence occurs in two stages: first we construct the simplest forms of sentence, and then, by means of reiterable devices of sentence-composition, we construct complex sentences out of them. The atomic sentences are to be thought of as made up out of a simple predicate, with an arbitrary but fixed number of argument-places, and a corresponding number of singular terms (the singular terms may themselves be complex, but I shall not concern myself with their internal structure). The modes of sentence-composition are twofold: by means of sentential operators, which may be unary, like negation, or binary, like the connectives 'or' and 'if'; and quantification, the device used in the formalized language for expressing generality. Quantification, as it is now generally considered, has two principal forms, universal ('for every *x*, . . .') and existential ('for some *x*, . . .'): the quantified sentence is formed by attaching one or other quantifier to a complex predicate. A complex predicate is, essentially, the result of removing one or more occurrences of some one singular term from a sentence; the quantified sentence must indicate where the resulting gaps occurred in the complex predicate. The sentence from which the complex predicate was formed may be regarded as a constituent of the quantified sentence; but, since we cannot determine, from the quantified sentence, any unique such constituent, we shall take as being a constituent of the quantified sentence any sentence from which the complex predicate could have been formed, that is, any sentence resulting from putting one and the same singular term in the gaps occurring in the complex predicate. (I have here described only first-order quantification; we shall not concern ourselves with quantification of higher order.)

It was in devising this syntax, so familiar in modern logic, and, in particular, in constructing the quantifier notation, that Frege's genius is principally revealed. Once you have a language whose syntax is of this form, it is not so difficult to go on to provide it with a semantics. I do not mean that Frege's semantics, now known as classical or two-valued semantics, is uncontentious; on the contrary, some logicians, for instance the intuitionists, reject it: only that it is not so hard to think of once you have the syntax. The first fundamental idea is that the condition for a complex sentence to be true depends solely upon its composition out of atomic sentences; here the conception under which each instance of a universally or existentially quantified sentence is taken to be a constituent of it is essential. *This* idea is not seriously challenged, at least by anyone accepting a Fregean syntax. Let us now define the *semantic value* of an atomic sentence to be whatever feature of it is both necessary and sufficient that it possess if every complex sentence is to be determined as true or otherwise in accordance with its composition out of atomic sentences. Then the second fundamental principle of Fregean, i.e. of classical, semantics is that the semantic value of an atomic sentence (and hence of any sentence) is just its truth-value—its being true or not true. Given the quantifier notation, it is not hard to come to the conclusion that the truth-value of a quantified sentence depends just on the truth-values of the constituents. It is less obvious that this is the case for a sentence formed by means of the connective 'if'; but it is no longer in dispute that, whether or not the 'if' of natural language works in this way, it is adequate for the formulation of mathematical and scientific theories to employ an 'if' for which the two-valued truth-table is correct. It is, nevertheless, the second principle to which those who reject classical semantics object.

Once we have these two principles, we can extend the notion of semantic value to expressions other than sentences; and there is no serious choice about how this extension is to be made. The semantic value of any expression is, again, that feature of it which must be ascribed to it if every sentence in which it occurs is to be determined as true or otherwise. By the second fundamental principle, the semantic value of a singular term or predicate depends solely on how an atomic sentence containing it is to be determined as true or false. We first lay down that the semantic value of a singular term is the particular object to which it refers: it then follows that the semantic value of a unary predicate is given by fixing of which objects it is true, of a binary predicate of which pairs of objects it is true, etc.

The account which I have just given is defective in an important respect: it depends upon assuming that, for every object, the language contains a singular term referring to it. Many languages will not satisfy this condition. For such a language, we must first specify the domain of quantification, and then consider an expansion of the language which contains a name of every object in that domain. If the domain is very large (non-denumerable), such an expanded language will only be an abstract construction, not a language that we can actually speak; but that does not matter, since we are using it only as an auxiliary device to give the semantics of the original language.

Once we have such a semantics, we can substitute for our notion of an interpretation by replacement that of a semantic interpretation, under which we make a direct assignment to the schematic letters of the semantic values of expressions of the appropriate categories, bypassing the expressions themselves.

The sense of an expression has been explained as that ingredient of its meaning which is relevant to the determination as true or false of a sentence in which it occurs: we can now say that it is that ingredient of its meaning which determines the semantic value of the expression. It is important here to avoid being confused over a point about which many philosophers express themselves ambiguously. One might say that the meaning of a sentence cannot, by itself, determine its truth or falsity, at least in the general case; hence the sense of the words can determine only the *condition* for its truth, not its truth-value; that will depend also upon how the world is, i.e., upon extra-linguistic reality. But the semantic value of an expression was so explained that the semantic values of the words composing a sentence will together determine it as true or false: when it is said that the sense of a word determines its semantic value, this is on the assumption that the contribution of extra-linguistic reality is being taken into account. The possibility of so explaining the notion of semantic value depends upon the assumption, embodied in the second fundamental principle of classical semantics, that the condition for the truth of each sentence is, determinately, either fulfilled or unfulfilled. We can regard this as a metaphysical asumption—an assumption of the existence of an objective reality independent of our knowledge. We can, equally, regard it as an assumption in the theory of meaning, namely that we succeed in conferring on our sentences a sense which renders them determinately true or false. Once this assumption is rejected, as it is for mathematical statements by the intuitionists, we can no longer legitimately explain the semantic value of an expression as going to determine each sentence in which it occurs as true or otherwise, since we cannot suppose the truth-value of every such sentence to be determined. On such a view, the semantic value of a sentence can, at best, be its truth-condition, and the conception of the semantic values of other expressions must be modified accordingly.

Classical semantics may be rejected by those who think that there are forms of sentence-composition, not included in Frege's formalized language, in whose presence it cannot be sustained; an example is the advocacy of a semantics of possible worlds by those who believe it necessary to account for modal operators like 'necessarily'. Or, more radically, it may be rejected even for a language using only Frege's forms of sentence-composition, by those who deny that every sentence so formed must be determinately true or false.

Frege's notion of the *reference* of an expression is, essentially, that of semantic value, as I have used this term, understood in the context of classical semantics. There is, indeed, another feature of Frege's notion with which I shall not be concerned, namely his belief that the semantic value of every expression can be given by associating with it some one extra-linguistic entity, of a logical type depending on the logical category of the expression; thus there is supposed to be some one thing whose association with a predicate determines, for each object, whether or not the predicate is true of that object. I shall pay almost no attention to this point.

Sense is an ingredient in meaning: to give an account of the sense of an expression is, therefore, to give a partial account of what a speaker knows when he understands that expression. In view of what I already said, that, in regarding sense as determining reference, we are supposing that the contribution of extra-linguistic reality is thereby taken into account, what Frege regarded as one of his fundamental discoveries, that there is a distinction between sense and reference, that is, that the sense of an expression cannot consist just in its having whatever reference it has, should be perfectly obvious: it is therefore at first sight surprising that it remains one of those of his theses which is most persistently controverted.

How is it even possible for anyone to reject the distinction between reference and sense? No one can deny that a capable speaker of the language must know more than the reference of a *complex* expression: he must at least know how its reference is determined in accordance with its composition out of its component words. Hence the most that can be maintained is that the distinction fails for single words, i.e. logically simple expressions; and the way is then clear, perhaps, for an admission that a speaker may understand a complex expression without knowing its reference—he knows the references of the component words, and knows how they jointly determine the reference of the whole, but does not actually know what that is. Indeed, this must be conceded unless it is to be held that any one who understands a sentence knows whether it is true or false. The difficulty is that an expression which is linguistically simple may have a complex sense: that is, its sense may be given as being that of some complex expression, as when we explain 'is prime' to mean 'has exactly two divisors' or 'liar' to mean 'man who asserts what he knows not to be true'. Hence either the application of the thesis that there is no distinction between sense and reference must be restricted still further, or it must be denied that, when we introduce a word by such means, we really do transfer to it the sense, as well as the reference, of the complex expression.

The attack on the sense/reference distinction comes from three directions: from the school of Davidson, according to which a theory of meaning for a language must take the form of a theory of truth in the style of Tarski; from the proponents of a holistic view of language, captained by Quine; and from the adherents of the so-called causal theory of reference, led by Kripke. Tarski was concerned with the question how we could convert the informal classical semantics, introduced by Frege for a formalized language, into a formal definition of truth for the sentences of that language: we are therefore now concerned with *two* formalized languages, rather than only one—the object-language for which we are defining truth, and the metalanguage in which we are defining it. For Frege, the reference of an expression is an extra-linguistic entity, and, in the informal semantics, or model theory, which has been developed from his ideas, an interpretation associates with each individual constant, predicate, etc., of the language a non-linguistic entity of a suitable type. When we are seeking to construct a truth-definition, however, attention shifts from the entities associated with expressions of the object-language as their references to the means available in the metalanguage of specifying this association. Moreover, Tarski, like Frege, but unlike model theorists in general, is concerned, not with the notion of truth under an arbitrary interpretation, but with truth simpliciter, that is, with truth under some one fixed interpretation (the intended one). In order to have a criterion for the success of a definition of truth for a given object-language, Tarski lays down a requirement which can be most simply stated for the case in which the metalanguage is an expansion of the object-language, namely that the definition be such that we shall be able to prove each instance of the schema

S is true if and only if *P*

which is obtained by putting for '*S*' a term denoting any sentence of the object-language and for '*P*' that sentence. In order to define 'true', it is necessary to define certain auxiliary notions, such as 'satisfies' and 'denotes'; and, in order to be able to prove each instance of the above schema, it is simplest to give to the clauses of their definitions the most direct form, that is, one in which the expression of the object-language which is mentioned is also used, as in

'*O*' denotes *O*

and

(u, v) satisfies 'x is greater than y' if and only if u is greater than v.

How much light is thrown on the analysis of the concept of truth by such a definition of 'true' is not our concern; Davidson has proposed that, if we reformulate it as an axiomatic theory rather than a definition, we may, by taking the notion of truth as already given, reconstrue it as giving a theory of meaning for the object-language. That is, whether or not actual speakers of the language can be credited with knowing such a theory, a knowledge of it by someone who possessed the requisite grasp of the concept of truth would confer on him an understanding of the language. Of course, if we are interested in the sort of knowledge exemplified by a person's mastery of his mother-tongue, this knowledge cannot consist in a capacity to give a verbal formulation of the theory of truth; we are interested in *what* he knows, not in his expression of that knowledge. In evaluating such a claim, we shall naturally enquire what is involved in attributing to a speaker a knowledge of such a theory, a knowledge that will not, in general, be explicit knowledge. I believe that Frege's arguments for the distinction between sense and reference already contain a demonstration of the inadequacy of such an account.

Table

		(C1) X knows the reference of w.
	(B2) X knows what is F.	(C2) X knows what w refers to.
	(B3) For some u, X knows, of u, that it is F.	(C3) For some u, X knows, of u, that w refers to it.
	(B4) X knows, of a, that it is F.	(C4) X knows, of a, that w refers to it.
(A5) X knows that P.	(B5) X knows that b is F.	(C5) X knows that w refers to b.
(A6) X knows that S is true.	(B6) X knows that $\ulcorner b$ is $F\urcorner$ is true.	(C6) X knows that $\ulcorner w$ refers to $b\urcorner$ is true.

Frege gave two such arguments, of which I shall offer a reconstruction rather than a direct account. What we are aiming to show is that to grasp the sense of a simple expression cannot be equated with knowing its reference, so we naturally wish to know how to understand an instance of the schema (C1) in the table. The table is arranged so that, e.g. (B5) is a special case of (A5), and (C5) a special case of (B5), and so on whenever there is more than one entry in any one row. Now it is natural to say that (C1) is equivalent to (C2); and so we wish to know how to understand instances of (B2). I shall say that an instance of (B2) ascribes 'knowledge-what' to the subject, as opposed to the 'knowledge-that' which is ascribed by an instance of (A5) obtained by replacing '*P*' by a complete sentence that might be used independently. Knowledge-what comprises also knowledge-which, knowledge-where, knowledge-when, and knowledge-who; it does not comprise knowledge-how in Ryle's sense, e.g. knowing how to mend a puncture, but does include such things as knowing how the poison was administered. An instance of (B2) is thus 'The police know who murdered Bexley'. A plausible suggestion is that (B2) is equivalent to (B3) (and thus (C1) and (C2) to (C3)). An instance of (B3) is an existential quantification of an instance of (B4), and thus will be true only if some corresponding instance of (B4) is true: e.g. 'The police know, of Redmayne, that he murdered Bexley'. I shall also speak of an instance of (A5) as ascribing *propositional knowledge* to the subject, and, in particular, as ascribing to him knowledge of the proposition expressed by the sentence which replaces '*P*'; by analogy, I shall say that an instance of (B4) ascribes *predicative knowledge* to the subject.

I should here interpolate the remark that the equation of (C1) with (C2) will be strictly correct only when there is some one entity which serves as the reference of the word in question, which Frege thought was always the case. When, e.g., '*w*' is replaced by a term referring to a unary predicate, we can, without invoking Frege's special doctrine, only say that someone knows the reference of the predicate when he knows of which objects it is true, and this statement is not exactly of the form (B2); we should get an instance of (B2) only by saying something like 'he knows which concept it refers to'. However, even if we do not accept Frege's special doctrine, the succeeding remarks can be transposed for cases other than that in which the word in question is a singular term.

We need now to know under what conditions an ascription of predicative knowledge (B4) will be true. It is natural to propose that the truth of an instance of (B5) will sometimes entail the truth of a corresponding instance of (B4), where '*a*' and '*b*' are replaced by expressions with the same reference, sometimes by the very same expression. For instance, it might be the case that the truth of 'The police know, of Redmayne, that he murdered Bexley' follows from the truth of 'The police know that Redmayne murdered Bexley'. Whenever this happens, I shall say that the ascription of predicative knowledge *rests on* an ascription of propositional knowledge. Thus the truth of certain instances of (B5) will ensure the truth of the corresponding instance of (B3), and hence of (B2). Of course, not every instance of (B5) will do this: the police may know that the man who hid on the roof murdered Bexley, but this will not be a ground for saying that they know who murdered Bexley unless we can independently say that they know who hid on the roof.

The suspicion may now arise that knowledge-what is not a sharp notion, or, at any rate, not one admitting a uniform characterization: for, it may be said, which pieces of propositional knowledge will validate an ascription of knowledge-what will depend upon the context. The police, for instance, want to know just enough about the murderer to be able to arrest and charge him, although there may be some sense in which they do not know who he (really) is. This is a plausible contention; the question is whether it is seen as applying equally to predicative knowledge, or whether it is held to be a ground for calling my equation of (B2) with (B3) in question. I shall assume, for the sake of argument, that predicative knowledge is a sharp and context-free notion; that, in so far as knowledge-what is not such a notion, this only shows that the equation of (B2) with (B3) is not wholly correct; but that, nevertheless, the equation of (C2) with (C3) *is* correct.

To ascribe to someone a knowledge of the reference of a word is, then, on this account, to assert the truth of some sentence of the form (C4); and sometimes this may be justified by appeal to the truth of a sentence of the form (C5), though we have so far not tried to circumscribe those sentences of that form which can serve this purpose. What, then, is it to ascribe to someone a *bare* knowledge of the reference of a word? It is, evidently, to ascribe to him a piece of predicative knowledge (C4), and to add that this is a *complete* characterization of this piece of knowledge on his part.

With this machinery, Frege's first argument for the distinction of sense and reference can be stated very simply. It has two premisses: first, that all theoretical knowledge is propositional knowledge, i.e. that every piece of predicative knowledge rests on some piece of propositional knowledge; and, secondly, that, for a given piece of predicative knowledge, there will never be a unique proposition knowledge of which will imply possession of that piece of predicative knowledge. It follows from these two premisses that there can be no such thing as a bare knowledge of the reference of a word, since the ascription of a given piece of predicative knowledge will never completely characterize the knowledge which the subject has: this can always be further characterized by citing the particular proposition on his knowledge of which his having that piece of predicative knowledge rests, and, when the knowledge is knowledge of the reference of a word, that further characterization displays the *sense* which the speaker attaches to the word.

This could be countered by denying either premiss. The first premiss seems to me certainly true; I do not believe there is any way to explain predicative knowledge save in terms of propositional knowledge: but I shall not pursue this question here. The second premiss is more uncertain, because of the unclarity of the notion of predicative knowledge. To answer someone who rejects this premiss therefore requires new considerations. Suppose that we replace '*w*' by an expression referring to a proper name, say the name 'Valencia': then it is hard to see what could be claimed as the unique replacement of '*b*' in (C5) which would justify (C2) save the name 'Valencia' itself; and this is just what is claimed by members of the Davidsonian school. This amounts, therefore, to claiming that to know what 'Valencia' refers to is just to know that 'Valencia' refers to Valencia. This claim seems on the face of it absurd, since it appears that anyone who knows that 'Valencia' is a proper name and that there is something to which it refers must know that 'Valencia' refers to Valencia, even if he does not know in the least what 'Valencia' is the name of: the proposal appears to conflate knowing the reference of a word with knowing that it has a reference. This obvious objection can be met, however, by invoking yet another distinction: that between knowing the truth of a sentence, as expressed by an instance of (A6), and knowing the proposition expressed by that sentence, as expressed by an instance of (A5). If someone knows that a certain sentence is true, it is a necessary condition for that knowledge to issue in knowledge of the proposition expressed by it that he should fully understand it (and thereby know what proposition it expresses). For instance, if someone who has never heard of Professor Geach or of semantics, and who knows of the word 'Germany' only that it is the name of some place in Europe, hears it authoritatively stated that Professor Geach is taking part in a conference on semantics to be held in Germany on 28 May, no one would say that he now knows that Professor Geach is participating in a conference on semantics in Germany on 28 May. He would know only that a certain sentence was true; more exactly, he might say that he knew that someone called 'Professor Geach' was taking part in a conference on something called 'semantics' to be held on 28 May in

Germany, wherever that was, thereby indicating of which words in the sentence he had an imperfect understanding.

If the distinction between (C6) and (C5) is to be invoked to defend the thesis that to know the reference of 'Valencia' is to know that 'Valencia' refers to Valencia, we must be told what is required for someone to know the proposition expressed by a sentence. It is certainly not necessary that he understand that sentence; he might both understand and know to be true an equivalent sentence in another language. If, for instance, the sentence contains the word 'sheep', he need not understand that particular English word, but it is necessary that he should grasp the concept it expresses; similarly, if the sentence contains the name 'Germany', he does not need to know the use of that name, but he must either know the use of some name, such as 'Deutschland' or 'Allemagne', having the same use, or at least have the conception of that use, i.e., of that way of picking out a specific geographical area, even if he does not associate it with any actual name in his vocabulary. Our hypothetical objector to Frege's argument cannot deny that there is something associated with a name which the subject must grasp if he is to be credited with knowledge of the proposition expressed by a sentence in which the name is used, without losing the distinction, for such a sentence, between knowing that proposition and knowing the sentence to be true; if he loses this, he is without defence against the charge that he has confused knowing the reference of a name with knowing that it has a reference. That which the subject must grasp is, precisely, the sense of the name.

The repudiation of the distinction between sense and reference, for proper names, is a manifestation of a tendency to revert to a doctrine of the kind advocated by Mill, that the linguistic function of a proper name consists in a direct association between the name and the object, of which no further account can be given. Frege's opponent may, therefore, claim that, while a grasp of a *complex* singular term—say 'the city besieged by the Cid'—will involve a particular way of picking out the object, a grasp of a proper name (more exactly, of a given *use* of a proper name) will not involve any such thing, but will consist in a direct association of the name with the object. What, then, is it to make such a mental association of a name with an object? The obvious temptation is to retreat to appealing once more to the notion of predicative knowledge—to say that it consists in knowing, of the object, that the name refers to it. But now it no longer matters whether either of the two premisses of Frege's argument holds good or not: whether or not all predicative knowledge rests on propositional knowledge, and whether or not, if so, there is, for each piece of predicative knowledge, more than one proposition a knowledge of which would entail it. Even if either premiss fails, it will remain that, to avoid circularity, the account that is being offered in place of Frege's must be supplemented by an explanation of what it is to have a piece of predicative knowledge which does not invoke the notion of propositional knowledge: and it was precisely the apparent impossibility of giving any such explanation that formed the ground for the first of the two premisses.

The only way to break the circle is to admit—what members of the Davidsonian school are reluctant to do—that to discuss meaning solely in terms of *what* a speaker knows when he grasps the use of a word is unsatisfactory. A theory of meaning, in the present sense, gives a theoretical representation of a practical ability; in Ryle's terms, it represents knowledge-how by knowledge-that. In many contexts, no great problem arises about what is involved in attributing to someone a piece of propositional knowledge, since he may be assumed to know some language in which the proposition can be expressed, and there will be no substantive difference between his knowing the proposition and his knowing the truth of the sentence; equally, there will be little difficulty about representing a piece of practical knowledge as knowledge of some proposition, since the transition between the two will be mediated by the subject's mastery of the language. But, when the knowledge with which we are concerned is itself that required for mastery of a language, these matters no longer take care of themselves. If we represent a speaker's understanding of a word as his knowledge of a certain proposition, our account remains incomplete unless we also explain in what an implicit knowledge of that proposition consists, that is, what is to count as a manifestation of that knowledge.

In these terms, what someone who knows the proposition expressed by a sentence containing a proper name must have is not, on this account, familiarity with a particular way of picking out or identifying the object, but simply an ability to recognize it when presented with it. We must not ask *how* or *by what* the object is recognized; even if there is an answer, the subject does not have to know it. Our understanding of the word 'red' depends on our ability to recognize something as red; we cannot answer the question *how* we recognize it.

Kripke has labelled Frege's thesis that proper names have sense 'the description theory', identifying it with the view that every proper name has the same sense as some definite description. It is, indeed, essential to Frege's view that a name *can* have the same sense as a definite description; but to think that a name can have no other kind of sense is seriously to misinterpret Frege. The idea that someone may have a capacity for recognizing an object which he cannot further explain is in no way absurd, and it would be quite wrong to suppose that Frege had any motive to deny that a grasp of a name might, on occasion, consist in its association with just such a capacity; although, conversely, the suggestion that our understanding of all, or even of many, names is of this kind is no more than ludicrous. But, of course, although a subject may be unable to give any *account* of such a capacity, it requires further *description*. We have, first, to ask what counts as the object's being 'presented' to him. Suppose the name is that of a city, say 'Valencia': is it being said that he can pick it out on a map, or that he can recognize it from afar, from the ground or from an aeroplane, or only that he can tell whenever he is in the city? If the latter, under what conditions?—only when he walks about the streets, or even when he is indoors, and, if indoors, only when he can see out of the window, or even when he is blindfolded? There is no one uniform notion of being *presented* with an object: a capacity to recognize it is relative to the conditions under which recognition takes place. Secondly—a point, due to Frege, of immense importance—his capacity to recognize the object must be subject to an awareness that it falls under the concept which determines the appropriate criterion for its identity (in our example, the concept of a city); in particular, he must be prepared to admit error when his claims to have recognized the object on specific occasions can be shown to contravene the criterion for its identity. When the ascription to the subject of such a capacity for recognition has been filled out in both these ways, there is no longer any ground in principle for denying that his grasp of the use of the name can consist in his ability so to identify its bearer, although this is no longer the simple matter it first seemed. Just that would be, in such a case, what constituted the sense which he attached to the name; there would be no force to the suggestion that, in giving an account of this sense, we should be saying no more than is conveyed by the bare statement that he knows, of the city, that a certain name refers to it.

Frege's first argument for the distinction between sense and reference has, however, a major defect: it has no tendency to show that the sense of a word is a feature of the *language*. It shows, at best, that each speaker, if he is to associate a reference with a word, must attach a particular sense to it; it does not show any necessity for different speakers to attach the same sense to any one word, so long as the senses which they all attach to it determine the same reference. It therefore leaves open the possibility that the sense of a word is not part of its meaning at all, if meaning is to be something objective and shared by all speakers, as Frege maintained; that, just as Frege said of the mental image associated with a word, it is merely a psychological accompaniment, or, at best, part of the psychological mechanism by which a speaker attaches a meaning to the word, and not a genuine ingredient of the meaning. This doubt is countered by Frege's second argument for the sense/reference distinction.

Frege's first argument says that we must ascribe *more* to a speaker than just a knowledge of the reference of a word; the second says that we cannot ascribe to him *as much*. This is the argument whereby Frege introduced the notion of sense, in his famous essay 'On Sense and Reference'.[1] The first argument related to a speaker's knowledge of the reference of a word, or, more generally, to what he knows when he knows its meaning. The second argument also has to do with knowledge, this time the knowledge conveyed to a hearer by an assertion, or, more generally, acquired by a speaker when he comes to accept a sentence as true. For a sentence to be able to be used to convey information, it must be possible to understand it in advance of knowing it to be true. Frege considers, specifically, an identity-statement, that is, one of the form '*a* is the same as *b*', and argues that if, in order to understand the statement, the hearer is required to know the reference of the terms '*a*' and '*b*', then, if it is true, he must, if he understands it, already know that it is true; hence, on this supposition, such a statement could convey no information. Plainly, this argument depends upon assuming that, if one knows the references of two terms, one must know whether or not they have the same reference. This assumption is obviously not justified under all possible ways of construing the expression 'to know the reference', e.g. not under one according to which the truth of an instance of (C1) follows from that of an instance of (C5), for the same replacement of '*w*' and *any* replacement of '*b*'. It can be justified by strengthening the thesis which Frege was refuting in either of two ways. First, let us take it as the thesis that to know the meaning of a term in part consists in (and does not merely entail) knowing its reference. It is an undeniable feature of the notion of meaning—obscure as that notion is—that meaning is *transparent* in the sense that, if someone attaches a meaning to each of two words, he must know whether these meanings are the same. It therefore follows from the thesis in this form that one who knows the meanings of two terms knows whether they have the same reference. Alternatively, we may state the thesis to be refuted as being that, if a

[1] Now Essay I of this volume.

speaker understands a term, he must have a *bare* knowledge of its reference, in the sense previously explained. The conclusion will then follow provided that we can say that, if someone knows, of an object u, both that it is F and that it is G, he must know, of that object, that it is both F and G. Again, this will obviously not hold good under every possible way of assigning the condition for someone to have a piece of predicative knowledge; but it will follow if we add the premiss that the ascription to him of those two pieces of predicative knowledge is a complete characterization of the relevant bits of knowledge on his part, since otherwise there would be nothing we could cite as the further piece of knowledge he would have to acquire in order to pass from the two pieces of knowledge that he had to the knowledge, of u, that it was both F and G.

Sense was defined as an ingredient in meaning, and consequently as being something that is transparent in the way that meaning is transparent, and as being that ingredient which goes to determine reference. If, then, it is necessary, for an identity-statement to be informative, that someone who grasps the senses of two terms may not know that they have the same reference, it follows that senses may differ although the reference is the same. Whether, now, we choose to say that someone who knows the sense also knows the reference is a matter of how we choose to construe the expression 'know the reference', i.e. the conditions we impose for the truth of an ascription of predicative knowledge. It seems less misleading *not* to say this: but, if we do say it, we can do so only because we allow a knowledge of the reference of any given term to rest on a knowledge of any of several distinct propositions, the difference between the propositions reflecting the difference between various senses that might be attached to the term without altering its reference.

This argument might seem to be a mere variation on the first one, employing a more circuitous route to the same conclusion; and, indeed, it is closely connected with it—I did not spell it out in complete detail, since that would have involved going over much of the same ground as with the first argument. The second argument has, however, a major advantage, in that it provides a reason for supposing that different speakers must attach the same sense to a word. For this second argument concerns the use of language for *communication*, which depends upon the informational content of a sentence being constant from speaker to speaker. If language is to serve as a medium of communication, it is not sufficient that a sentence should in fact be true under the interpretation placed on it by one speaker just in case it is true under that placed on it by another; it is also necessary that both speakers should be aware of the fact.

The notion of informational content is plainly connected, in a manner which I shall not here attempt to make precise, with that of what we recognize as justifying a statement, as establishing its truth: a statement with no informational content requires no justification, and it is plausible to say that we grasp the information that is conveyed by a sentence just in case we know what is needed to establish it as true. It is therefore plain that a parallel argument for the distinction between sense and reference can be constructed in terms of what is required for us to recognize an identity-statement as true. Such an argument would equally provide a reason for taking the sense attached to a term as common to different speakers, since the successful use of language depends upon agreement over what is needed to show a statement to be true. There is, indeed, a problem here for a theory of meaning, like Frege's, which issues in a classical, two-valued semantics. Not every sentence has a sense such that we are able, even in principle, to recognize it as true, provided only that it is true. It follows that the senses of certain expressions cannot be given solely in terms of our means of recognizing as true sentences in which they occur; rather, they must be given in such a way that we grasp what condition must obtain for such a sentence to be true, independently, at least in part, of whether we are able to recognize that condition as obtaining. An explanation is therefore called for of how we know, for a sentence of this kind, what to count as showing that it is true in a case in which we are able to recognize this, since the sense of the sentence was not directly given by reference to our means of recognition. It is plain that, if a theory of meaning of a Fregean kind is to be capable of accounting satisfactorily for the actual use of language, such an explanation must be forthcoming; the detailed account of the senses of our words must be such as to display the means by which we are able to derive, from a grasp of the condition which must hold for a sentence to be true, a knowledge of when we may recognize that condition as fulfilled.

Frege's second argument for the sense/reference distinction may be generalized to other forms of sentence, in fact to any atomic sentence. To know the reference of a predicate is to know which objects it is true of; that is, if knowledge-which is to be construed as predicative knowledge, to know, of each object in the domain, whether or not the predicate is true of it. Hence, if someone knows the reference of a predicate, and also knows, of some object, that it is what a given term refers to, then he must know whether or not the sentence which results from attaching that predicate to that term is true. This argument exactly parallels Frege's argument about identity-statements, and is valid under the corresponding way of construing the thesis shown to lead to absurdity.

There is, however, a difference between Frege's argument and this extension of it. To say that, in general, the semantic value of a singular term is an object does not in any way restrict the kind of semantic theory we adopt (unless some particular ontology, some doctrine about the kinds of object which the world contains, is presupposed); hence Frege's original argument does not depend upon the adoption of a classical, two-valued semantics. But the extension of the argument to other atomic sentences depends upon assuming that the semantic value of a predicate is its extension, i.e. its being determinately true or false of each object in the domain; and this holds good only within classical semantics.

But, just as the assumption underlying the extension of the argument is stronger, so the conclusion is more powerful. Frege's argument about identity-statements would be met by supposing the sense of a singular term to be related to its reference as a programme to its execution, that is, if the sense provided an effective procedure of physical and mental operations whereby the reference could be determined. For, without appeal to the assumption that every meaningful sentence *has* a determinate truth-value, we cannot claim that the semantic value of each sentence is its truth-value; and hence, although we may say that it is possible to understand a sentence without knowing its semantic value, we do not have a ground for arguing that one may understand it without even being able effectively to discover its semantic value. Thus, granted that the semantic value of a singular term is the object to which it refers, we cannot, without appeal to the principle of bivalence, assume that there can be admissible singular terms whose reference cannot be effectively determined, nor, therefore, any identity-statements whose truth-value cannot be effectively decided. The extension of the argument to all atomic sentences does presuppose bivalence: hence it leaves open the possibility that the language may contain primitive predicates whose application cannot be decided effectively, and which we can therefore understand without being able to determine their semantic value, since the semantic value of a predicate is being assumed to be its extension. As already remarked, some distinction between sense and semantic value must be admitted whatever semantic theory we adopt; but it is only in the context of classical semantics that we require a notion of sense which determines the semantic value of an expression, but in a non-effective manner.

My discussion of the notion of sense has been programmatic only, just as Frege's was: I have not said specifically what we must take as constituting the senses of expressions of different categories, but have merely discussed the arguments for supposing that we need such a notion. My own opinion is that it is highly dubious that classical semantics is correct; if it is not, then we need admit only a comparatively trivial distinction between sense and semantic value. However, in order to investigate this exceedingly deep question, we need to have an accurate conception of the form which a theory of meaning embodying classical semantics should take; and my conviction is that such a theory must incorporate a substantial distinction between reference and sense of just the kind that Frege wished to draw.

Frege's arguments for the distinction do not in themselves provide a defence against the attack on it by the holistic school. As a characteristic expression of holism, let us take the following passage from Davidson:

> To give up the analytic-synthetic distinction as basic to the understanding of language is to give up the idea that we can clearly distinguish between theory and language. Meaning, as we might loosely use the word, is contaminated by theory, by what is held to be true.[2]

I will expound the thought here expressed in terms of Quine's ideas rather than those of Davidson himself (an adequate treatment would, naturally, require examination of all the differing versions of holism). A detailed account of the way in which a specific language functions—what I have hitherto been calling a 'theory of meaning' for the language—is to be judged correct or incorrect solely by whether it is possible to derive from it the observable linguistic dispositions of the speakers. These dispositions may be considered simply as dispositions to assent to or dissent from sentences of the language under certain sensory stimuli, either unconditionally or conditionally upon prior assent to or dissent from other sentences. The conditional dispositions constitute the inferential connections between sentences which are acknowledged by the speakers. The linguistic dispositions possessed by speakers at a given time may include or involve a preparedness to accept certain sentences as true independently of current external stimulus. Suppose, now, that a new sentence comes to be generally accepted as true. This may happen for a variety of reasons: on the basis of observation, as the conclusion of a chain of reasoning, or simply by stipulation. However it happens, it will, in virtue of the inferential connections between the newly accepted sentence and other sentences, alter many of the speakers' conditional and unconditional dispositions to assent or dissent. It will make just the

[2] D. Davidson, 'On the Very Idea of a Conceptual Scheme', *Proceedings and Addresses of the American Philosophical Association*, 46 (1973–4), 5–20; see p. 9.

same alterations in them, whatever the reason for which it was originally accepted as true: in fact, there will be nothing in the subsequent linguistic behaviour of the speakers on the basis of which we could differentiate between this sentence and any other which they accept as true.

I have tried to state this doctrine as coolly as possible: but it is a heady doctrine, and has gone to the heads of many of its proponents. For it looks as though we can describe the upshot of the doctrine like this: Frege thought of language as a game played with fixed rules, there being all the difference in the world between a move in the game and an alteration of the rules; but, if holism is correct, every move in the game changes the rules—the distinction has no basis in reality. Some holists, indeed, have gone so far as to claim to have shown the notion of meaning itself to be superfluous; all have treated that of a change of meaning as useless. One reason why the opponents of holism have so far thrown so little light on the question is their failure to distinguish between the fundamental doctrine of holism, as I have just done my best to expound it, and the consequences which its proponents have claimed it to have. The fundamental doctrine is, in my opinion, at the best no more than a half-truth; but more important is the fact that, even if it is wholly true, it does not and cannot have the consequences claimed for it.

Let us imagine that we have, for a given language, a theory of meaning from which we can derive the current conditional and unconditional linguistic dispositions of the speakers. The theory will therefore provide for a speaker's making certain judgements in certain circumstances; it will, moreover, itself display, via the inferential connections encapsulated in the conditional dispositions, how a speaker's total set of dispositions will be modified by the subsequent acceptance or rejection of any given sentences. Hence, by the principles of the theory of meaning which we have, we shall not be able to determine a speaker's actual linguistic dispositions at future dates by appeal to the theory of meaning alone: we can do so only by this taken together with an accumulating stock of judgements that he accepts.

So far, we have assumed neither that the fundamental doctrine of holism applies to this language, nor that it does not: but suppose now that it does, that is, that there is no way to discriminate, by appeal to the linguistic behaviour of speakers at any given time, between any one sentence generally accepted as true and any other, in particular no way of discriminating between them according to the kind of reason which the speakers originally had for accepting them. Then there will be something arbitrary in the representation of a speaker's linguistic dispositions by means of the constant theory of meaning together with the accumulating stock of judgements: the details of the representation would depend upon where we happened to come in. If we had started the enterprise earlier or later, the stock of judgements would be, respectively, larger or smaller, and the fixed theory of meaning would be different. Hence, although our representation is not incorrect, we may prefer a different one which is not arbitrary in this way. A theory of meaning for the language of this new type would not, of itself, allow us to derive any specific linguistic dispositions at all: it would, instead, yield a function which mapped every set (or, perhaps, every finite set) of judgements as to the truth and falsity of sentences on to a total set of conditional and unconditional dispositions to assent and dissent. A theory of meaning of this kind would be uniform: its form would not depend upon the arbitrary selection of an origin; but it would remain the case that, given such a theory of meaning, we could determine the actual linguistic dispositions of a speaker at any time only by first knowing which sentences he accepted as true and which he rejected as false.

Just what, in detail, such a theory of meaning would look like, I have no idea; I do not accept the fundamental doctrine of holism, and so I do not need to know. What is clear is that, like any workable theory of meaning, it must have a finite base: the knowledge that a speaker has, in knowing the language, is a finite amount of knowledge, although it issues in the understanding of infinitely many sentences. Like any theory of meaning, its atomic ingredients would be the representations of the meanings of the individual words and forms of sentence-construction of the language: these, taken together, would collectively determine an effective mapping of (finite) sets of judgements on to total ranges of linguistic dispositions. One mistake often made by holists is to claim that, on a holistic theory, a sentence does not have a meaning of its own; this is as absurd as to say that, on a Fregean theory, a word does not have a sense of its own. What does follow from holism is that the theory of meaning for the language does not by itself determine our disposition to assent to or dissent from any one particular sentence under different conditions; but that does not entail that, on such a theory, the sentences and even the words of the language do not have a meaning.

It is, furthermore, plain that there is no plausible sense of 'meaning'—however loosely we employ the term—under which it may be claimed that the general acceptance of a judgement will effect a change of meaning, nor, therefore, one under which we may say that the theories embraced by the speakers will infect the meanings of their words. True enough, on *any* theory, the making of a judgement will modify the dispositions which a speaker has to assent to or dissent from other sentences; but there is no more ground to regard those dispositions as determinative of meaning on a holistic theory than on a Fregean one. The modification of the linguistic dispositions that occurs as a result of the making of a judgement takes place in accordance with the general understanding of the language which is embodied in an adequate theory of meaning for it, and so a change in these dispositions is no indication of any change in the meanings of the words. Languages do, indeed, change, in vocabulary, in pronunciation, in syntax, and, there is no reason to doubt, in meaning. But we need to acknowledge as a change in meaning only a change that does not happen according to rule, a change which cannot be accounted for by theory; perhaps we ought to say, that cannot be accounted for by a theory an implicit grasp of which can be attributed to every speaker, since, if the speakers are unable to foresee that a given change would occur in hypothetical circumstances, it cannot be the meanings of the words which determine that change, even if it is predictable. But the changes which a holist claims as consequent upon the making of any judgement, or at least upon its adoption by the whole linguistic community, and as being a change in meaning 'as we might loosely use the word', are not of this kind.

It is thus plain that a satisfactory theory of meaning for a language which satisfies the fundamental doctrine of holism would have to be of very much *greater* complexity than one for a language which does not: in constructing a theory of meaning for a language of the first kind, the identification of the linguistic dispositions currently possessed by speakers from a finite set of observations of their linguistic behaviour would be only a first step: the second step would be the much more difficult one of framing general principles for determining a total range of such dispositions for any (finite) set of sentences generally accepted as true. Certainly no one has come anywhere near even a sketch of what such a holistic theory of meaning would be like (Quine's *Word and Object* is concerned with translation and not with a theory of meaning). There is a tendency amongst holists to speak as though they had dispensed with the necessity for a theory of meaning altogether, as though, having, as they think, shown meaning to be contaminated by (speakers') theory, they had thereby shown that we need have no truck with the notion at all. This, if correct, would imply that all we need to be told is the theory which a speaker, or a linguistic community, holds at a given time—the set of sentences accepted as true—and we shall thereby be able to understand both those sentences and any others in the language, the distinction between an interpreted and an uninterpreted theory having fallen away as untenable. This is a complete illusion. Given simply a bunch of sentences accepted as true by all speakers, or by any one speaker, there is no way at all to arrive at any unique interpretation of those sentences, or of other sentences of the same language. The contrary impression is, I suppose, arrived at in some such way as this: suppose we consider a set of sentences corresponding to those which all contemporary speakers of English would consider true, save that throughout 'table' replaced 'eagle' and vice versa; then the information that those were the sentences considered as true by the speakers of some language would suffice to tell us what, in that language, the words 'table' and 'eagle' meant (as we might loosely say); for the set would include such sentences as 'Eagles, when standing on their legs, have flat horizontal upper surfaces', 'Female tables lay eggs', etc., etc. But, of course, we are here tacitly appealing to the assumption that we already know the linguistic dispositions of speakers which relate to sentences not containing the words 'table' and 'eagle': it trades on the supposition that, against this background, we can display the inferential connections between sentences containing those words and other sentences simply by indicating which sentences containing them are taken by everybody to be true, and that, against the same background, these inferential connections will suffice to determine the unconditional dispositions to assent to or dissent from sentences containing them. It has no tendency at all to show that there is any means of deriving, just from the set of generally accepted sentences, the linguistic dispositions of the speakers. Holism is not a doctrine which allows us to abandon the idea that individual words have senses; on the contrary, it is one which, if correct, demands that we regard our words as having senses of a much more complex kind than we have imagined, of a kind, indeed, of which we have as yet formed no clear picture.

It might be objected that I have relied on a simple denial of the thesis for which Davidson has contended, that, in constructing a theory of meaning for a language, we have nothing to go on, at the outset, but a knowledge of the sentences which the speakers hold true and the conditions obtaining when they make such judgements. Davidson is, however, thinking of a language which, like all natural languages, contains a great many indexical devices, and the judgements which provide the data include many judgements concerning sentences containing such devices. The fundamental doctrine of holism, as I formulated it, does not apply to such sentences, which are, of course, judged true only as uttered on a particular occasion; no one could maintain that no feature of our linguistic behaviour will discriminate between such sentences and those claimed by an adherent of a Fregean theory of meaning to be analytic.

Moreover, Davidson's data include information about the conditions prevalent when a judgement was originally made, so that, although Davidson is himself a holist, he is taking as part of his data what, in the case of a sentence without indexical features which comes to be generally accepted, Quine declares to be an irrelevant piece of historical information. Davidson's thesis is totally different from that which I am denying. What I am denying is that one can derive, from the knowledge that a certain set of sentences—necessarily ones without indexical features—comprises all those accepted as true by all speakers of a language, without any further information about the conditions leading to their acceptance, the linguistic dispositions of the speakers, or anything that could possibly be taken as an interpretation of the language. I do not say that any holist has ever explicitly advanced this thesis: but it is the thesis that would have to hold if the fundamental doctrine of holism were to be a ground for saying that we can dispense with the notion of meaning or that that notion has been engulfed by that of speakers' theory; and it is patently false.

For instance, let us come down to earth and ask what happens, in the context of a holistic view of language, to the conception of the sense of a proper name. It is common for holists to respond to this by saying that, since we no longer regard any generally accepted sentence containing the name as having a privileged status, if we are to retain the notion of sense at all, it must be taken as given by the totality of all sentences containing the name which are generally accepted as true; there cannot be a selection, among such sentences, of those which are constitutive of the sense of the name, i.e., go to determine its reference, as opposed to those which merely record what we believe to be true of its bearer. This may seem reasonably plausible for names of objects not or no longer accessible to observation, though implausible for other names: but it does not give the sense as this would have to be presented in a genuinely holistic theory of meaning of the kind I have been discussing, for this would be a sense which did not determine the truth of any particular sentences containing the name, but would map any given set of such sentences on to appropriate linguistic dispositions. On the contrary, this explanation of sense by the holist represents the supplanting of meaning by the speakers' theory, and will change as that theory changes. The suggestion in fact exemplifies in a very striking way the justice of my claim that holists tend to be under the illusion that they have shown how we may do without any theory of meaning, whereas they have merely made it very much harder to see how to construct one.

But, it may be said, all that this shows is that an unrestricted application of the fundamental doctrine of holism would lead to a theory of meaning of unmanageable complexity: it does not show that that doctrine is not rightly to be applied to certain sentences, e.g. to those that might be used in telling someone the reference of a proper name, nor that, so applied, it would have any such awkward consequences. Is it not plausible in itself that we can make no effective distinction between the sense of a proper name and the totality of what is generally believed about its bearer? And, if so, is it not illegitimate to object against such a doctrine that we could not simultaneously apply such a thesis to words of all kinds?

Undoubtedly proper names provide the most difficult case for the distinction between sense and reference, which is why discussions of that distinction tend to concentrate upon them; but such discussions diminish in value according to the difficulty of generalizing them to other words; and, in the plea I have just enunciated, such generalization has been forsworn. There is, indeed, some plausibility in the contention: but, to test it, we must appeal again to our distinction between knowing that a sentence is true and knowing the proposition expressed by it. If a child who has never heard of Milan sees a newspaper headline 'Postal strike in Milan', he does not know that the postmen are on strike in Milan, but only that the sentence 'The postmen are on strike in Milan' is true, or that the postmen are on strike in a place called 'Milan'. What would he have to learn for us to say of him that he knew the proposition? Not, surely, *everything* that is generally believed about Milan, e.g. that St Ambrose was its bishop, that first the Viscontis and then the Sforzas were its dukes, etc., etc. Even in the case of proper names, there is room for a distinction between the standard explanation of its reference and the provision of standard information about its bearer: only an obsessive adherence to a *theory* could give us any reason to seek to deny this.

The so-called causal theory of reference attacks the sense/reference distinction from the opposite direction: for it, what would ordinarily be said in explanation of a proper name has no role in determining the reference of that name, since it might prove false. For instance, the standard way to explain the name 'Edward Gibbon' to someone unfamiliar with it would be to say that Gibbon wrote a book called *The History of the Decline and Fall of the Roman Empire*; we might, nevertheless, discover that that book was not after all written by Gibbon. What determines the reference of a name is therefore taken to be nothing that is shown by the way in which the name might be explained, but the existence of a causal chain connecting a given use of a name with its original introduction. To avoid difficulties about a case in which one individual is named after another, it is then laid down that each particular utterance which forms a link in this causal chain must be made with the intention of preserving the reference of the name; but then further provisos have to be made to allow for the possibility that such an intention can sometimes be frustrated, that there may be an unwitting transfer of the name from one bearer to another; this, it is explained, can happen because, although there is an intention to preserve the reference, it is accompanied by some other over-riding intention. This possibility has been admitted for place-names (as with Evans's excellent example of 'Madagascar', originally the name of part of the mainland, transferred in error to the island); it is hard to see on what ground one could then rule out cases of transference of a person's name in his lifetime (for instance, of a changeling), or, indeed, after his death (it is far from evident, in Kripke's own example, that if in fact 'Goliath' was originally the name of another Philistine, and not of the giant whom David killed, we, in using the name to refer to the latter, are saying anything which we ought to withdraw). Once all these concessions are made, and given that it is agreed that the criterion for what determines the reference is what the speakers would accept as determining it in a case of conflict, the causal theory has shrunk from a rival to Frege's conception that the reference of a proper name is determined by its sense to a proposal, for a restricted class of proper names, about what should be taken as an important ingredient of their senses.

The causal theory, in a full-blown form, makes it impossible to distinguish between knowing the use of a proper name and simply having heard the name and recognizing it as a name, and hence between knowing a proposition and knowing the truth of the sentence expressing it, when this turns on knowing a name occurring in that sentence. What difference does it make that someone is able to give the standard explanation of the name 'Edward Gibbon' if that explanation may be factually incorrect without affecting his grasp of the name? Suppose I overhear someone say, 'Edward Gibbon was received into the Catholic Church', and that this is the first time I have heard the name, and know nothing of the topic of the conversation: if the causal theory is correct, I nevertheless implicitly understand what determines the reference of the name, and could use it myself with the same reference merely by intending to do so. In what way, then, do I fail to grasp the use of the name? In what way am I worse off than someone who can give the standard explanation, particularly if, as it happens, that standard explanation should be incorrect?

The difference is surely that if I use the name 'Edward Gibbon' knowing no more about its bearer than what I learned from that snatch of conversation, one would need to go back to the particular speakers from whom I overheard it to find out who was being talked about; but, if someone is in possession of the standard explanation of the name, then we should never consider it relevant to trace the particular way in which he first heard the name, even if it should be discovered that the standard explanation is wrong, i.e. that it was not Gibbon who wrote *The Decline and Fall*. It is for this reason that we should say that, in the first case, I did not know that Edward Gibbon was once received into the Catholic Church, but knew only that this was true of someone called 'Edward Gibbon'; even if it be granted that, in many cases, the standard explanation of a name is not to be taken as definitively true of its bearer, under pain of the name's being deprived of reference, still a knowledge of that standard explanation may be an essential ingredient of a grasp of the use of the name as a word of the common language. To whatever extent it is correct to say that the causal ancestry of a name determines its reference, it is the causal ancestry of that name in its generally agreed use that counts, and not the causal ancestry of any old utterance of the name made with the intention of preserving the reference, whatever that might have been. Kripke has argued, correctly, that the point that Gibbon might prove not to have been the author of *The Decline and Fall* cannot be met by declaring the sense of the name 'Edward Gibbon' to be 'the man generally believed to have written *The Decline and Fall*', since the general belief so appealed to is the belief that Edward Gibbon wrote *The Decline and Fall*, and to have such a belief presupposes a grasp of the name. But this should not obscure the fact that that whose causal ancestry we must trace is the belief, not the name itself.

What makes it possible, even for someone who knows nothing more of Gödel than that he proved the incompleteness of arithmetic, to entertain the supposition that it might turn out that it was not he who did so, is his awareness that there are other means, not in dispute, and discoverable on enquiry even though not presently known to him, of identifying the bearer of the name 'Gödel'. It is for that reason that the notion of knowing, of Gödel, that he proved the incompleteness of arithmetic has more substance than that of knowing that 'Gödel' is the name of the man who proved that theorem. The same, indeed, holds for the name 'Gibbon'. But even when a name retains an association with only one act, e.g. 'Obadiah' with the composition of the Old Testament prophecy, we may suppose that this tradition has come down to us from the lifetime of the author; and, since the original attribution from which that tradition stemmed could have been mistaken, but would have taken substance from the other means, then lying to hand, of referring to the individual in question, it is intelligible, and plausible, to hold that, even in such a case, the name on our lips refers to the supposed rather than the real author. But here it is the *tradition* which connects our use of the name with the man; where the actual *name* itself first came from has little to do with it. This is why a case like that of Goliath is quite different. Here the tradition has to do with the Philistine giant; it is not to be supposed that, at the time, anyone took a different man to be the one whom David had killed with his sling, but only that, then or at a later date, they got his *name* wrong: and so the reference of the name, as *we* use it, is to the man to whom the tradition relates, and not the one who bore that name when he lived.

The causal theory of reference embodies a genuine insight into the way the reference of a special class of proper names is determined: names of persons, animals, ships, etc., which were objects of acquaintance, i.e. possible objects of ostension, when their names were introduced, but which are no longer in existence. Even for these I have argued that the theory is not quite right as it stands, and that it is in any case only a theory *about* the senses of such names rather than one which *replaces* sense by something different. What it certainly does not do is to give an account of the functioning of proper names in general. We are of course especially interested in personal proper names, which is one reason why philosophers who discuss proper names tend to use personal names for most of their examples. But we should not forget that personal names have several features peculiar to them which are not shared by proper names of objects of other kinds, and that these are very various: place-names, names of months and days of the week, names of races (like 'the Grand National'), names of games, of chess openings, of stars and constellations, of mathematical theorems, of poems, kinds of dance, religions, winds, diseases, scripts and languages, hurricanes, exhibitions, wars and treaties, whirlpools, scientific theories. Of very few of these is it even intelligible to say that they might turn out to belie the means we should use to explain their names to someone who did not know them; and of very few kinds of name is it plausible to hold that their reference, as we use them, depends upon what the name was originally introduced for. It would, I believe, be a grave mistake to suppose that, with the causal theory, the mechanism of reference has at last been uncovered; still less that the notion of the sense of a name has been shown to be redundant.

XIV

WITTGENSTEIN ON FOLLOWING A RULE

JOHN McDOWELL

These things are finer spun than crude hands have any inkling of.
(*RFM* VII–57.)[1]

1

We find it natural to think of meaning and understanding in, as it were, contractual terms.[2] Our idea is that to learn the meaning of a word is to acquire an understanding that obliges us subsequently—if we have occasion to deploy the concept in question—to judge and speak in certain determinate ways, on pain of failure to obey the dictates of the meaning we have grasped; that we are 'committed to certain patterns of linguistic usage by the meanings we attach to expressions' (W, p. 21).[3]

From *Synthese*, 58 (1984), 325–63. Reprinted by permission of Kluwer Academic Publishers.
This essay originated in an attempt to respond to Simon Blackburn's 'Rule-Following and Moral Realism', in Steven Holtzman and Christopher Leich (eds.), *Wittgenstein: To Follow a Rule* (London: Routledge and Kegan Paul, 1981), 163–87; I was stimulated also, in writing the first draft, by an unpublished paper of Blackburn's called 'Rule-Following'. I have been greatly helped by comments on the first draft from Margaret Gilbert, Susan Hurley, Saul Kripke, David Lewis, Christopher Peacocke, Philip Pettit, David Wiggins, and Crispin Wright, who also kindly let me see a draft of his 'Kripke's Wittgenstein', a paper presented to the Seventh Wittgenstein Symposium at Kirchberg, Austria, in August 1982, and forthcoming in the *Journal of Philosophy*.

[1] I shall use '*RFM*' for the third edition of Wittgenstein's *Remarks on the Foundations of Mathematics*, ed. G. H. von Wright, R. Rhees, and G. E. M. Anscombe, and trans. G. E. M. Anscombe (Oxford: Blackwell, 1978).

[2] See p. 19 of Crispin Wright, *Wittgenstein on the Foundations of Mathematics* (London: Duckworth, 1980); hereafter referred to by 'W'.

[3] This idea of commitment to patterns must be treated with care if we are not to falsify the intuition. The most straightforward sort of case, on which it is familiar that Wittgenstein concentrates, is the continuation of a numerical series. Here it is natural to think of the correct expansion of the series as constituting a pattern to which understanding of its principle commits one. In the general case, the 'pattern' idea is the idea of a series of things that, given the way the world develops, it would be correct to say if one chose to express a given concept; outside the series-expansion case, this idea is obviously metaphorical at best, since what it is correct to say with the use of a given concept, even supposing a determinate state of

According to Crispin Wright, the burden of Wittgenstein's reflections on following a rule, in his later work, is that these natural ideas lack the substance we are inclined to credit them with: 'there is in our understanding of a concept no rigid, advance determination of what is to count as its correct application' (ibid.).[4]

If Wittgenstein's conclusion, as Wright interprets it, is allowed to stand, the most striking casualty is a familiar intuitive notion of objectivity. The idea at risk is the idea of things being thus and so anyway, whether or not we choose to investigate the matter in question, and whatever the outcome of any such investigation. That idea requires the conception of how things could correctly be said to be anyway—whatever, if anything, we in fact go on to say about the matter; and this notion of correctness can only be the notion of how the pattern of application that we grasp, when we come to understand the concept in question, extends, independently of the actual outcome of any investigation, to the relevant case. So if the notion of investigation-independent patterns of application is to be discarded, then so is the idea that things are, at least sometimes, thus and so anyway, independently of our ratifying the judgement that that is how they are. It seems fair to describe this extremely radical consequence as a kind of idealism.[5]

We may well hesitate to attribute such a doctrine to the philosopher who wrote:

> If one tried to advance theses in philosophy, it would never be possible to debate them, because everyone would agree to them. (*PI* §128.)[6]

Notice that the destructive effect of the doctrine goes far beyond Wittgenstein's hostility to the imagery of mathematical platonism, in which mathematics is pictured as 'the natural history of mathematical

affairs one aims to describe, depends on what other concepts one chooses to express in the same utterance. (The non-metaphorical kernel is simply the idea that the meaning of what one says is a matter of the conditions under which it would be true.) It is important, also, not to falsify the connection between the patterns and meaningfulness—for instance, by suggesting that the idea is that making sense depends on *conforming* to the appropriate commitments. Tracing out the patterns is what the 'pattern' idea takes consistently speaking the truth to be; to make sense (in an affirmation) one needs to do no more than felicitously make as if to be doing what one takes that to require. (See, further, n. 41 below.)

[4] 'Rigid' will call for comment: see n. 19 below.

[5] Wright does this at p. 252 of 'Strict Finitism', *Synthese*, 51 (1982). See also pp. 246–7 of his 'Anti-Realist Semantics: The Role of *Criteria*', in Godfrey Vesey (ed.), *Idealism: Past and Present* (Cambridge: CUP, 1982), 225–48.

[6] I shall use '*PI*' for *Philosophical Investigations*, trans. G. E. M. Anscombe (Oxford: Blackwell, 1953). Stanley Cavell's correction of the usual reading of this passage, at pp. 33–4 of his *Claim of Reason* (Oxford: Clarendon Press, 1979), does not make it any easier to reconcile with Wright's view of Wittgenstein.

objects' (*RFM* II-40). The remarks about rule-following are not confined to mathematics; on Wright's reading they would undermine our ordinary intuitive conception of natural history, literally so called—the very model on which that suspect platonist picture of mathematics is constructed.

More specific grounds for doubting the attribution might be derived from passages like this (*PI* §195):

> "But I don't mean that what I do now (in grasping a sense) determines the future use *causally* and as a matter of experience, but that in a *queer* way, the use itself is in some sense present."—But of course it is, 'in *some* sense'! Really the only thing wrong with what you say is the expression "in a queer way". The rest is all right. . . .[7]

What this suggests is something we might anyway have expected: that Wittgenstein's target is not the very idea that a present state of understanding embodies commitments with respect to the future, but rather a certain seductive misconception of that idea.

Not that Wright merely ignores such passages. His claim (see W, p. 21) is that Wittgenstein seems *almost* to want to deny all substance to the 'pattern' idea; what he attributes to Wittgenstein (see W, p. 227) is not an outright abandonment of the idea but a reinterpretation of it. Wright's view is that the intuitive contractual picture of meaning and understanding can be rendered innocuous—purged of the seductive misconception—by discarding the thought that the patterns are independent of our ratification. Later (Sections 5, 7, 10) I shall suggest that this purged version of the intuitive picture is not recognizable as a picture of meaning and understanding at all, and is not correctly attributed to Wittgenstein. But for the present, let me note only that Wright's reinterpretation, precisely by denying the ratification-independence of the patterns, leaves the intuitive conception of objectivity untenable, in the way I described above. So we are bound to wonder whether the concession that Wright envisages Wittgenstein making to the 'pattern' idea can account satisfactorily for Wittgenstein's reassuring tone in his response to the interlocutor of *PI* §195.

2

In Wright's view, then, the butt of Wittgenstein's reflections on rule-following is the idea that understanding an expression is 'grasp of a

[7] See also e.g. *PI* §§187, 692, 693.

pattern of application, conformity to which requires certain determinate verdicts in so far unconsidered cases' (W, p. 216). But:

We have to acknowledge . . . that the 'pattern' is, strictly, inaccessible to definitive explanation. For, as Wittgenstein never wearied of reminding himself, no explanation of the use of an expression is proof against misunderstanding; verbal explanations require correct understanding of the vocabulary in which they are couched, and samples are open to an inexhaustible variety of interpretations. So we move towards the idea that understanding an expression is a kind of 'cottoning on'; that is, a leap, an inspired guess at the pattern of application which the instructor is trying to get across. (W, p. 216.)

The pictured upshot of this 'leap' is something idiolectic. So the suggestion is that the 'pattern' idea comes naturally to us, in the first instance, in the shape of 'the idea that each of us has some sort of privileged access to the character of his own understanding of an expression; each of us knows of an idiolectic pattern of use, for which there is a strong presumption, when sufficient evidence has accumulated, that it is shared communally' (W, p. 217).[8]

What is wrong with this idea? Wright's answer is this:

whatever sincere applications I make of a particular expression, when I have paid due heed to the situation, will seem to me to conform with my understanding of it. There is no scope for a distinction here between the fact of an application's seeming to me to conform with the way in which I understand it and the fact of its really doing so.[9]

Now we are naturally inclined to protect the intuitive view that thoughts and utterances make sense by virtue of owing, or purporting to owe, allegiance to conceptual commitments. So, given that idiolectic understanding cannot make room for the 'pattern' idea, it is tempting to appeal to communal understanding. But (the argument that Wright ascribes to Wittgenstein continues) this cannot rehabilitate the 'pattern' idea. For (W, p. 218):

Suppose that one of us finds himself incorrigibly out of line concerning the description of a new case. We have just seen that he cannot single-handed, as it were, give sense to the idea that he is at least being faithful to his *own* pattern; that is, that he recognises how he must describe the new case if he is to remain faithful to his own understanding of the relevant expressions. How, then, does his disposition to apply the expression to a new case become, properly speaking, recognition of the continuation of a pattern if it so happens that he is *not* out of line, if it so happens that there is communal agreement?

[8] See also W, pp. 32, 354.

[9] See also W, p. 36. Compare *PI* §258: 'One would like to say: whatever is going to seem right to me is right. And that only means that here we can't talk about "right".' (See Section 14 below.)

The trouble is that there is a precise parallel between the community's supposed grasp of the patterns that it has communally committed itself to and the individual's supposed grasp of his idiolectic commitments. Whatever applications of an expression secure communal approval, just those applications will seem to the community to conform with its understanding of the expression.[10] If we regard an individual as aiming to speak a communal language, we take account of the possibility that he may go out of step with his fellows; thus we make room for an application of the notion of error, and so of right and wrong. But it is only going out of step with one's fellows that we make room for; not going out of step with a ratification-independent pattern that they follow. So the notion of right and wrong that we have made room for is at best a thin surrogate for what would be required by the intuitive notion of objectivity. That would require the idea of concepts as authoritative; and the move away from idiolects has not reinstated that idea. In sum (W, p. 220):

> None of us unilaterally can make sense of the idea of correct employment of language save by reference to the authority of securable communal assent on the matter; and for the community itself there is no authority, so no standard to meet.

3

According to Wright, then, Wittgenstein's reflections are directed, in the first instance, against the idea that a determinate practice can be dictated by a personal understanding—something that owes no allegiance to a communal way of going on. On the surface, at least, there is a point of contact here with Saul Kripke's influential reading of the remarks on rule-following, which I shall now outline.[11]

Suppose one is asked to perform an addition other than any one has encountered before, either in the training that gave one one's understanding of addition or in subsequently trying to put one's understanding into practice.[12] In confidently giving a particular answer, one will naturally have a thought that is problematic: namely—to put it in terms

[10] One would like to say: whatever is going to seem right to *us* is right. And that only means that here we can't talk about 'right'.

[11] See Saul A. Kripke, 'Wittgenstein on Rules and Private Language: An Elementary Exposition', in Irving Block (ed.), *Perspectives on the Philosophy of Wittgenstein* (Oxford: Blackwell, 1981), 238–312; hereafter referred to by 'K'. Wright notes the point of contact at p. 249 of 'Strict Finitism'; though he takes issue with Kripke in 'Kripke's Wittgenstein'.

[12] Where I say 'other', Kripke has 'larger'. This makes the scepticism perhaps more gripping, but the difference is inessential.

that bring out the point of contact with Wright's reading—that in returning this answer one is keeping faith with one's understanding of the 'plus' sign. To show how this thought is problematic, Kripke introduces a sceptic who questions it. The natural idea is that one's understanding of 'plus' dictates the answer one gives. But what could constitute one's being in such a state? Not a disposition: no doubt it is true that answering as one does is an exercise of a disposition that one acquired when one learned arithmetic, but the relation of a disposition to its exercises is in no sense contractual—a disposition is not something to which its exercises are faithful.[13] But nothing else will serve either: for—to quote Kripke's summary of a rich battery of argument—'it seems that no matter what is in my mind at a given time, I am free in the future to interpret it in different ways' (K, p. 294). That is, whatever piece of mental furniture I cite, acquired by me as a result of my training in arithmetic, it is open to the sceptic to point out that my present performance keeps faith with it only on one interpretation of it, and other interpretations are possible. So it cannot constitute my understanding 'plus' in such a way as to dictate the answer I give. Such a state of understanding would require not just the original item but also my having put the right interpretation on it. But what could constitute my having put the right interpretation on some mental item? And now the argument can evidently be repeated.

The upshot of this argument is a 'sceptical paradox', which, according to Kripke, Wittgenstein accepts: there is no fact that could constitute my having attached one rather than another meaning to the 'plus' sign (K, pp. 272–3).

It may well seem that if Wittgenstein concedes this much to Kripke's sceptic, he has renounced the right to attribute meaning to expressions at all. According to Kripke, however, Wittgenstein offers a 'sceptical solution' to the 'sceptical paradox'. (A 'sceptical solution' to a sceptical problem is one that 'begins . . . by conceding that the sceptic's negative assertions are unanswerable' (K, p. 270).) The essentials of this 'sceptical solution' are as follows.

First, we must reform our intuitive conception of meaning, replacing the notion of truth conditions with some notion like that of justification conditions. Kripke quotes with approval (K, p. 274) a claim of Michael Dummett's: 'The *Investigations* contains implicitly a rejection of the classical (realist) Frege–*Tractatus* view that the general form of explana-

[13] This is the gist of the excellent discussion at K, pp. 250–7.

tion of meaning is a statement of the truth conditions.'[14] The 'sceptical paradox', which we are to accept, is that there is no fact that could constitute my having attached one rather than another determinate meaning to the 'plus' sign. We are inclined to understand this as a concession that I have attached *no* determinate meaning to the 'plus' sign: but the suggestion is that this is only because we adhere, naively, to the superseded truth-conditional conception of meaning—applied, in this case, to the claim 'I have attached a determinate meaning to the "plus" sign'. (See K, p. 276.)

Second, when we consider the justification conditions of the statements in which we express the idea that someone attaches some determinate meaning to an expression (the conditions under which we affirm such statements, and the roles they play in our lives), we see that we can make sense of them in terms of their use to record acceptance of individuals into the linguistic community. (The thesis that we can make sense of the idea of meaning only in that connection is the core of Kripke's interpretation of the Private Language Argument.)

Now there is room for doubt about how successful this 'sceptical solution' can be. The exegetical framework within which it is constructed—the Dummettian picture of the transition between the *Tractatus* and the *Investigations*—is not beyond dispute. But without opening that issue (which I shall touch on below: Sections 10, 11, 14), we can note that when Dummett expresses his doubts about the 'realist' (truth-conditional) conception of meaning (which are supposed to be in the spirit of the later Wittgenstein's doubts about the *Tractatus*), it is typically by pressing such questions as this: 'What could constitute someone's possession of the sort of understanding of a sentence that "realism" attributes to him?' The implication is that, failing a satisfactory answer, no one could possess that sort of understanding.[15] It is natural to suppose that if one says 'There is no fact that could constitute its being the case that P', one precludes oneself from affirming that P; and this supposition, so far from being a distinctively 'realist' one, plays a central role in the standard arguments *against* 'realism'. Given this supposition, the concession that Kripke says Wittgenstein makes to the sceptic becomes a *denial* that I understand the 'plus' sign to mean one thing rather than another. And now—generalizing the denial—we do seem to have fallen into an abyss: 'the incredible and self-defeating conclusion, that all language is

[14] 'Wittgenstein's Philosophy of Mathematics', *Philosophical Review*, 68 (1959), 324–48, at p. 348.

[15] See especially Michael Dummett, 'What is a Theory of Meaning? (II)', in Gareth Evans and John McDowell (eds.), *Truth and Meaning* (Oxford: Clarendon Press, 1976), 67–137.

meaningless' (K, p. 273). It is quite obscure how we could hope to claw ourselves back by manipulating the notion of accredited membership in a linguistic community.

4

In any case, Kripke's thesis that Wittgenstein accepts the 'sceptical paradox' seems a falsification. Kripke (see K, p. 241) identifies the 'sceptical paradox' that he attributes to Wittgenstein with the paradox that Wittgenstein formulates in the first paragraph of *PI* §201:

> This was our paradox: no course of action could be determined by a rule, because every course of action can be made out to accord with the rule. The answer was: if everything can be made out to accord with the rule, then it can also be made out to conflict with it. And so there would be neither accord nor conflict here.

But §201 goes on with a passage for which Kripke's reading makes no room:

> It can be seen that there is a misunderstanding here from the mere fact that in the course of our argument we give one interpretation after another; as if each one contented us at least for a moment, until we thought of yet another standing behind it. What this shows is that there is a way of grasping a rule which is *not* an *interpretation*, but which is exhibited in what we call "obeying the rule" and "going against it" in actual cases.

What could constitute my understanding, say, the 'plus' sign in a way with which only certain answers to given addition problems would accord? Confronted with such questions, we tend to be enticed into looking for a fact that would constitute my having put an appropriate *interpretation* on what I was told and shown when I was instructed in arithmetic. Anything we hit on as satisfying that specification contents us only 'for a moment'; then it occurs to us that whatever we have hit on would itself be capable of interpretation in such a way that acting in conformity with it would require something quite different. So we look for something that would constitute my having interpreted the first item in the right way. Anything we come up with as satisfying that specification will in turn content us only 'for a moment'; and so on: 'any interpretation still hangs in the air along with what it interprets, and cannot give it any support' (*PI* §198). Kripke's reading has Wittgenstein endorsing this reasoning, and consequently willing to abandon the idea that there is anything that constitutes my understanding an expression in some determinate way. But what Wittgenstein clearly claims, in the second paragraph of §201, is that the reasoning is vitiated by 'a misunderstanding'. The right response to the paradox, Wittgenstein in effect tells us, is not to accept it but to correct the misunderstanding on which it depends: that is, to realize 'that there is a way of grasping a rule which is *not* an *interpretation*'.

The paradox of §201 is one horn of a dilemma with which the misunderstanding presents us. Suppose we are not disabused of the misunderstanding—that is, we take it that our problem is to find a fact that constitutes my having given some expression an interpretation with which only certain uses of it would conform. In that case, the attempt to resist the paradox of §201 will drive us to embrace a familiar mythology of meaning and understanding, and this is the second horn of the dilemma. My coming to mean the expression in the way I do (my 'grasping the rule') must be my arriving at an interpretation; but it must be an interpretation that is not susceptible to the movement of thought in the sceptical line of reasoning—not such as to content us only until we think of another interpretation standing behind it.

> What one wants to say is: "Every sign is capable of interpretation; but the *meaning* mustn't be capable of interpretation. It is the last interpretation." (*Blue Book*, p. 34.)[16]

Understanding an expression, then, must be possessing an interpretation that cannot be interpreted—an interpretation that precisely bridges the gap, exploited in the sceptical argument, between the instruction one received in learning the expression and the use one goes on to make of it. The irresistible upshot of this is that we picture following a rule as the operation of a super-rigid yet (or perhaps we should say 'hence') ethereal machine.

> How queer: It looks as if a physical (mechanical) form of guidance could misfire and let in something unforeseen, but not a rule! As if a rule were, so to speak, the only reliable form of guidance. (*Zettel* §296.)[17]

One of Wittgenstein's main concerns is clearly to cast doubt on this mythology. But his attacks on the mythology are not, as Kripke suggests, arguments for acceptance of the 'sceptical paradox'.[18] That would be so

[16] *The Blue and Brown Books* (Oxford: Blackwell, 1958). Compare *Zettel*, ed. G. E. M. Anscombe and G. H. von Wright, and trans. G. E. M. Anscombe (Oxford: Blackwell, 1967), §231.

[17] There is a good description of the mythological ideas expressed here, with a wealth of citations of relevant passages, in Gordon Baker, 'Following Wittgenstein: Some Signposts for *Philosophical Investigations* §§143–242', in Holtzman and Leich, *To Follow a Rule*, 31–71.

[18] See K, pp. 269, 272: Kripke cannot distinguish rejection of the 'superlative fact' of *PI* §192—rejection of the mythology—from refusing to countenance a fact in which my attaching a determinate meaning to 'plus' consists—acceptance of the paradox.

if the dilemma were compulsory; but the point of the second paragraph of *PI* §201 is precisely that it is not. The mythology is wrung from us, in our need to avoid the paradox of the first paragraph, only because we fall into the misunderstanding; the attack on the mythology is not support for the paradox, but rather constitutes, in conjunction with the fact that the paradox is intolerable, an argument against the misunderstanding.

It is worth noting two points about the second horn of the dilemma that correspond to two aspects of Wright's reading of Wittgenstein.

First, if we picture an interpretation that would precisely bridge the gap between instruction and competent use, it seems that it can only be one which each person hits on for himself—so that it is at best a fortunate contingency if his interpretation coincides with the one arrived at by someone else subjected to the same instruction, or with the one intended by the instructor.

"But do you really explain to the other person what you yourself understand? Don't you get him to *guess* the essential thing? You give him examples,—but he has to guess their drift, to guess your intention." (*PI* §210.)

This is clearly the basis in Wittgenstein for Wright's remarks (quoted in Section 2 above) about 'the idea that understanding an expression is a kind of "cottoning on"; that is, a leap, an inspired guess at the pattern of application which the instructor is trying to get across' (W, p. 216).

Second, a concomitant of the picture of the super-rigid machine is a picture of the patterns as sets of rails. (See, for instance, *PI* §218.) At each stage, say in the extending of a series, the rule itself determines what comes next, independently of the techniques that we learn in learning to extend it; the point of the learning is to get our practice of judging and speaking in line with the rule's impersonal dictates. (An omniscient God would not need to do mathematics in order to know whether '777' occurs in the decimal expansion of π; see *RFM* VII-41.) Now this conception figures regularly in Wright's formulations of the 'pattern' idea:

. . . the pattern extends *of itself* to cases which we have yet to confront . . .

. . . the investigation-independent truth of statements requires that their truth is settled, *autonomously and without the need for human interference*, by their meanings and the character of the relevant facts.[19]

[19] W, p. 216, and 'Strict Finitism', p. 250; both with my emphasis. 'Rigid', at W, p. 21 (quoted in Section 1 above), is an expression of the same idea—Wright does not mean 'rigid' as opposed to, say, 'vague' (see Baker, 'Following Wittgenstein', pp. 40–1).

It is clear, again, that these formulations have a basis in Wittgenstein's polemic against the second horn of the dilemma. A remark like 'I give the rule an extension' (*RFM* VI-29) is meant as a corrective of the inclination to say 'The rule extends of itself'. (And 'even God can determine something mathematical only by mathematics': *RFM* VII-41.)

5

In Wright's reading, as I said (Sections 1 and 2 above), Wittgenstein's point is that the natural contractual conception of understanding should not be discarded, but purged of the idea—which it must incorporate if the intuitive notion of objectivity is to have application—that the patterns to which our concepts oblige us are ratification-independent. I expressed a suspicion (in Section 1 above) that this purging would not leave a residue recognizable as a conception of meaning and understanding at all, or recognizable as something that Wittgenstein recommends. I want now to begin on an attempt to back up this suspicion.

At *PI* §437 Wittgenstein writes:

> A wish seems already to know what will or would satisfy it; a proposition, a thought, what makes it true—even when that thing is not there at all! Whence this *determining* of what is not yet there? This despotic demand? ("The hardness of the logical must.")

Note the parenthesis: clearly he thinks that the discussion in which this passage occurs—dealing with the relation between wishes or expectations and their fulfilment, and the relation between orders and their execution—raises the same issues as his reflections on the continuation of a series. (See K, p. 300, n. 17.) We can bring out the connection by focusing on the case of orders and their execution: it is natural to say that the execution of an order is faithful to its meaning, and in saying this we clearly express a version of the idea that we express when we say that the competent continuation of a series is faithful to its principle.

What would Wright's reading of Wittgenstein be like, transposed to this case? Something on these lines (cf. Section 2 above). The temptation to say that my execution of an order conforms with my understanding of it arises primarily out of a conception of my understanding as idiolectic—something that cannot be definitively conveyed to someone else, so that it is at best a happy contingency if it coincides with the understanding of the order possessed by the person who issued it. On reflection, however, we should realize that this is an illusion: we cannot make sense of anything that would constitute an essentially personal understanding of an order, but would nevertheless impose genuine constraints on what I did in 'execution' of it. For whatever I 'sincerely' did would seem to be in conformity with my supposed personal understanding of the order. We naturally want to protect the intuitive notion of an action's fulfilling an order; so we are tempted at this point to appeal to the idea of my membership in a linguistic community. This does make room for my going wrong. But all that my going wrong can amount to is this: my action does not secure the approval of my fellows, or is not what they would do in attempted fulfilment of such an order. When the community does approve, that is not a matter of its collectively recognizing the conformity of my action to an antecedent communal understanding of the order: for this supposed communal understanding would be in exactly the same position as my supposed idiolectic understanding. We cannot hold, then, that the community 'goes right or wrong', by the lights of its understanding, when it awards my action the title 'execution of the order'; 'rather, it just goes' (W, p. 220).

Given the correspondence (noted in Section 4 above) between aspects of Wright's reading and aspects of Wittgenstein's polemic against the second horn of the dilemma, it is not surprising that part, at least, of this transposed version of Wright's reading should neatly fit parts of Wittgenstein's discussion. Consider, for instance, *PI* §460:

> Could the justification of an action as fulfilment of an order run like this: "You said 'Bring me a yellow flower', upon which this one gave me a feeling of satisfaction; that is why I have brought it"? Wouldn't one have to reply: "But I didn't set you to bring me the flower which should give you that sort of feeling after what I said!"?

It seems correct and illuminating to understand this as an attack on the idea that the understanding I act on is essentially idiolectic.[20]

Taken as a whole, however, I think this reading gets Wittgenstein completely wrong. I can perhaps begin to explain my disbelief with this remark: it would have been fully in character for Wittgenstein to have written as follows:

> Could the justification of an action as fulfilment of an order run like this: "You said 'Bring me a yellow flower', upon which this one received approval from all the bystanders; that is why I have brought it"? Wouldn't one have to reply: "But

[20] That is, the passage is of a piece with the passage from *PI* §258 quoted in n. 9 above. This suggestion does not compete with, but rather complements, Kripke's suggestion (K, p. 300, n. 17) that the passage refers obliquely to Russell's treatment of desire in *The Analysis of Mind* (London: George Allen and Unwin, 1921).

I didn't set you to bring the flower which should receive approval from everyone else after what I said!"?

In his later work, Wittgenstein returns again to trying to characterize the relation between meaning and consensus. If there is anything that emerges clearly, it is that it would be a serious error, in his view, not to make a radical distinction between the significance of, say, 'This is yellow' and the significance of, say, 'This would be called "yellow" by (most) speakers of English' (see, for instance, *Zettel* §§ 428–31). And my transposed version of Wright's reading seems to leave it mysterious, at best, why this distinction should be so important.

It may appear that the answer is both obvious and readily available to Wright: 'To say "This would be called 'yellow' by speakers of English" would not be to *call* the object in question "yellow", and that is what one does when one says "This is yellow".' But this would merely postpone the serious question: does Wright's reading of Wittgenstein contain the means to make it intelligible that there should so much as *be* such an action as calling an object 'yellow'? The picture Wright offers is, at the basic level, a picture of human beings vocalizing in certain ways in response to objects, with this behaviour (no doubt) accompanied by such 'inner' phenomena as feelings of constraint, or convictions of the rightness of what they are saying. There are presumably correspondences in the propensities of fellow members of a linguistic community to vocalize, and to feel comfortable in doing so, which are unsurprising in the light of their belonging to a single species, together with similarities in the training that gave them the propensities. But at the basic level there is no question of shared commitments—of the behaviour, and the associated aspects of the streams of consciousness, being subject to the authority of anything outside themselves. ('For the community itself there is no authority, so no standard to meet': W, p. 220.) How, then, can we be entitled to view the behaviour as involving, say, calling things 'yellow', rather than a mere brute meaningless sounding off?

The thought that is operative here is one that Kripke puts by saying: 'The relation of meaning and intention to future action is *normative, not descriptive*' (K, p. 257). It is a thought that Wright aims to respect. This is the point of his aspiration not to discard the contractual conception of meaning, but only to purge it of the idea of ratification-independence. But the purging yields the picture of what I have been calling 'the basic level'; and at that level Wright's picture has no room for norms, and hence—given the normativeness of meaning—no room for meaning. Wright hopes to preserve a foothold for a purified form of the normativeness implicit in the contractual conception of meaning, by appealing to the fact that individuals are susceptible to communal correction. It is problematic, however, whether the picture of the basic level, once entertained as such, can be prevented from purporting to contain *the real truth* about linguistic behaviour. In that case its freedom from norms will preclude our attributing any genuine substance to the etiolated normativeness that Wright hopes to preserve. The problem for Wright is to distinguish the position that he attributes to Wittgenstein from one according to which the possibility of going out of step with our fellows gives us the *illusion* of being subject to norms, and consequently the *illusion* of entertaining and expressing meanings.

6

Moved by the insight that meaning relates normatively to linguistic behaviour, Kripke—like Wright—reads Wittgenstein as concerned to preserve a role for the intuitive contractual conception. But Kripke's Wittgenstein locates that conception only in the context of the 'sceptical solution'—a response to a supposedly accepted 'sceptical paradox'. Applied to the case of orders and their execution, Kripke's 'sceptical paradox' will take this form: there is nothing that constitutes my understanding an order in a way with which only acting in a certain determinate manner would conform. And, here as before (cf. Section 4 above), it is open to question whether, once that much is conceded to scepticism, a 'sceptical solution' can avert the destructive effect that the concession threatens to have.

In any case, this line of interpretation gets off on the wrong foot, when it credits Wittgenstein with acceptance of a 'sceptical paradox', so that a 'sceptical solution' would be the best that could be hoped for. Just as in the case of the continuation of a series, the reasoning that would lead to this 'sceptical paradox' starts with something that Wittgenstein aims to show up as a mistake: the assumption, in this case, that the understanding on which I act when I obey an order must be an interpretation. The connection with the thought of *PI* §201 is made clear by this juxtaposition (*RFM* VI-38):

> How can the word "Slab" indicate what I have to do, when after all I can bring any action into accord with any interpretation?
>
> How can I follow a rule, when after all whatever I do can be interpreted as following it?

The parallel can be extended (see Section 4 above). If we assume that understanding is always interpretation, then the need to resist the paradox of *PI* §201 drives us into a fantastic picture of how understanding mediates between order and execution. Consider, for instance, *PI* §431:

> "There is a gulf between an order and its execution. It has to be filled by the act of understanding."
> "Only in the act of understanding is it meant that we are to do THIS. The order—why, that is nothing but sounds, ink-marks.—"[21]

The act of understanding, conceived in terms of hitting on an interpretation that completely bridges the gulf between an order and its execution, demands to be pictured as setting up a super-rigid connection between the words and the subsequent action (hence the allusion, in *PI* §437, to 'the hardness of the logical must'). It is this idea that Wittgenstein is mocking in *PI* §461:

> In what sense does an order anticipate its execution? By ordering *just that* which later on is carried out?—But one would have to say "which later on is carried out, or again is not carried out." And that is to say nothing.
> "But even if my wish does not determine what is going to be the case, still it does so to speak determine the theme of a fact, whether the fact fulfils the wish or not." We are—as it were—surprised, not at anyone's knowing the future, but at his being able to prophesy at all (right or wrong).
> As if the mere prophecy, no matter whether true or false, foreshadowed the future; whereas it knows nothing of the future and cannot know less than nothing.

And the parallel goes further still. When we are tempted to conceive the understanding of an order in this way, what we have in mind is something essentially personal: a guess at the meaning of the person who issued the order. This idea is Wittgenstein's target in, for instance, *PI* §433:

> When we give an order, it can look as if the ultimate thing sought by the order has to remain unexpressed, as there is always a gulf between an order and its execution. Say I want someone to make a particular movement, say to raise his arm. To make it quite clear, I do the movement. This picture seems unambiguous until we ask: how does he know *he is to make this movement*?—How does he know at all what use he is to make of the signs I give him, whatever they are?—Perhaps I shall now try to supplement the order by means of further signs, by pointing from myself to him, making encouraging gestures, etc. Here it looks as if the order were beginning to stammer.
> As if the signs were precariously trying to produce understanding in us.—But if we now understand them, by what token do we understand?

If we read Wittgenstein in Kripke's way, we shall take Wittgenstein's mockery of these ideas as argument in favour of the 'sceptical

[21] Compare the passage from *Blue Book*, p. 34, quoted in Section 4 above.

paradox'—the thesis that there is nothing that could constitute my understanding an order in a determinate way. That is what the mockery would amount to if there were no options besides the paradox and the ideas that Wittgenstein mocks. But Wittgenstein's point is that this dilemma seems compulsory only on the assumption that understanding is always interpretation; his aim is not to shift us from one horn of the dilemma to the other, but to persuade us to reject the dilemma by discarding the assumption on which it depends.

7

Having diagnosed the dilemma as resting on the mistaken idea that grasping a rule is always an interpretation, Wittgenstein goes on, famously, to say (*PI* §202):

> And hence also 'obeying a rule' is a practice. And to *think* one is obeying a rule is not to obey a rule. Hence it is not possible to obey a rule 'privately': otherwise thinking one was obeying a rule would be the same thing as obeying it.

The diagnosis prompts the question 'How can there be a way of grasping a rule which is not an interpretation?', and I think the thesis that obeying a rule is a practice is meant to constitute the answer to this question. That is, what mediates the inference ('hence also') is this thought: we have to realize that obeying a rule is a practice if we are to find it intelligible that there is a way of grasping a rule which is not an interpretation. (The rest of §202—the crystallization into two sentences of the Private Language Argument—is offered as a corollary.)

There is another formulation of the same line of thought in *PI* §198:

> "Then can whatever I do be brought into accord with the rule?"—Let me ask this: what has the expression of a rule—say a sign-post—got to do with my actions? What sort of connexion is there here?—Well, perhaps this one: I have been trained to react to this sign in a particular way, and now I do so react to it.
>
> "But that is only to give a causal connexion: to tell how it has come about that we go by the sign-post; not what this going-by-the-sign really consists in."—On the contrary, I have further indicated that a person goes by a sign-post only in so far as there exists a regular use of sign-posts, a custom.[22]

This passage opens with an expression of the paradox formulated in the first paragraph of §201. Then Wittgenstein introduces the case of sign-posts, in order to adumbrate the diagnosis that he is going to state more

[22] I have ventured to change the punctuation in the second paragraph, in order to make the dialectical structure of the passage clearer.

explicitly in §201. When I follow a sign-post, the connection between it and my action is not mediated by an interpretation of sign-posts that I acquired when I was trained in their use. I simply act as I have been trained to.[23] This prompts an objection, which might be paraphrased on these lines: 'Nothing in what you have said shows that what you have described is a case of following a rule; you have only told us how to give a causal explanation of certain bits of (what might as well be for all that you have said) mere behaviour.' The reply—which corresponds to the first sentence of §202—is that the training in question is initiation into a custom. If it were not that, then the account of the connection between sign-post and action would indeed look like an account of nothing more than brute movement and its causal explanation; our picture would not contain the materials to entitle us to speak of following (going by) a sign-post.[24]

Now how exactly is this to be understood?

Wittgenstein's concern is to exorcize the insidious assumption that there must be an interpretation that mediates between an order, or the expression of a rule given in training, on the one hand, and an action in conformity with it, on the other. In his efforts to achieve this, he is led to say such things as 'I obey the rule *blindly*' (*PI* §219). This is of a piece with his repeated insistence that the agreement that is necessary for the notion of following a rule to be applicable is not agreement in opinions:

> "So you are saying that human agreement decides what is true and what is false?"—It is what human beings *say* that is true and false; and they agree in the *language* they use. That is not agreement in opinions but in form of life. (*PI* §241.)[25]

I take it that at least part of the point of this passage is that an opinion is something for which one may reasonably be asked for a justification; whereas what is at issue here is below that level—the 'bedrock' where 'I have exhausted the justifications' and 'my spade is turned' (*PI* §217). The thought is clear in *RFM* VI-28:

> Someone asks me: What is the colour of this flower? I answer: "red".—Are you absolutely sure? Yes, absolutely sure! But may I not have been deceived and called the wrong colour "red"? No. The certainty with which I call the colour "red" is the rigidity of my measuring-rod, it is the rigidity from which I start. When I give descriptions, *that* is not to be brought into doubt. This simply characterizes what we call describing.

[23] Compare *PI* §506: 'The absent-minded man who at the order "Right turn!" turns left, and then, clutching his forehead, says "Oh! right turn" and does a right turn.—What has struck him? An interpretation?'

[24] Compare *RFM* VI-43.

[25] See also *RFM* VI-30, VI-49.

(I may of course even here assume a slip of the tongue, but nothing else.)

Following according to the rule is FUNDAMENTAL to our language-game. It characterizes what we call description.

Again (*RFM* VI-35):

How do I know that the colour that I am now seeing is called "green"? Well, to confirm it I might ask other people, but if they did not agree with me, I should become totally confused and should perhaps take them or myself for crazy. That is to say: I should either no longer trust myself to judge, or no longer react to what they say as to a judgement.

If I am drowning and I shout "Help!", how do I know what the word Help means? Well, that's how I react in this situation.—Now *that* is how I know what "green" means as well and also know how I have to follow the rule in the particular case.[26]

What Wittgenstein is trying to describe is a use of language in which what one does is 'to use an expression without a justification' (*PI* §289; compare *RFM* VII-40). One may be tempted to protest: when I say 'This is green', in the sort of case he envisages, I do have a justification, namely that the thing in question is green. But how can I justify the use of an expression by repeating it? It is thoughts of this sort that lead Wittgenstein to say (*On Certainty* §204):

Giving grounds, however, justifying the evidence, comes to an end;—but the end is not certain propositions' striking us immediately as true, i.e. it is not a kind of *seeing* on our part; it is our *acting*, which lies at the bottom of the language-game.[27]

Now there is a temptation to understand this on the following lines. At the level of 'bedrock' (where justifications have come to an end), there is nothing but verbal behaviour and (no doubt) feelings of constraint. Presumably people's dispositions to behaviour and associated feelings match in interesting ways; but at this ground-floor level there is no question of shared commitments—everything normative fades out of the picture.

This is the picture of what I called 'the basic level' that is yielded, in Wright's reading, by the rejection of ratification-independence (see Section 5 above). I expressed disbelief that a position in which this is

[26] With 'Well, that's how I react in this situation', compare *PI* §217: 'I am inclined to say: "This is simply what I do."'

[27] *On Certainty*, ed. G. E. M. Anscombe and G. H. von Wright, and trans. G. E. M. Anscombe and Denis Paul (Oxford: Blackwell, 1969). It is worth noting how paradoxical 'it is not a kind of seeing' can seem in the case of such uses of language as saying that something is green. For an illuminating discussion of Wittgenstein's stress on acting as lying 'at the bottom of the language-game', see Peter Winch, 'Im Anfang war die Tat', in Block, *Perspectives*, pp. 159–78.

how things are at the basic level can accommodate meaning at all. If it is true that a failure to accommodate meaning is the upshot of the position, then it can be attributed to Wittgenstein only at the price of supposing that he does not succeed in his aims. But we are now equipped to see that the attribution falsifies his intentions. When he describes the 'bedrock' use of expressions as 'without justification', he nevertheless insists (to complete the quotation from *PI* §289): 'To use an expression without a justification does not mean to use it without right.'[28] And it seems clear that the point of this is precisely to prevent the leaching out of norms from our picture of 'bedrock'—from our picture, that is, of how things are at the deepest level at which we may sensibly contemplate the place of language in the world. To quote again from *RFM* VI-28: 'Following according to the rule is FUNDAMENTAL to our language-game.' By Wittgenstein's lights, it is a mistake to think we can dig down to a level at which we no longer have application for normative notions (like 'following according to the rule'). Wright's picture of the basic level, so far from capturing Wittgenstein's view, looks like a case of succumbing to a temptation that he is constantly warning against: 'The difficult thing here is not, to dig down to the ground; no, it is to recognize the ground that lies before us as the ground' (*RFM* VI-31). Wittgenstein's problem is to steer a course between a Scylla and a Charybdis. Scylla is the idea that understanding is always interpretation. This idea is disastrous because embracing it confronts us with the dilemma of Section 4 above: the choice between the paradox that there is no substance to meaning, on the one hand, and the fantastic mythology of the super-rigid machine, on the other. We can avoid Scylla by stressing that, say, calling something 'green' can be like crying 'Help!' when one is drowning—simply how one has learned to react to this situation. But then we risk steering on to Charybdis—the picture of a basic level at which there are no norms; if we embrace that, I have suggested, then we cannot prevent meaning from coming to seem an illusion. The point of *PI* §198, and part of the point of §§201–2, is that the key to finding the indispensable middle course is the idea of a custom or practice. How can a performance both be nothing but a 'blind' reaction to a situation, not an attempt to act on an interpretation (avoiding Scylla); and be a case of going by a rule (avoiding Charybdis)? The answer is: by belonging to a custom (*PI* §198), practice (*PI* §202), or institution (*RFM* VI-31).

[28] Or 'wrongfully' (*RFM* VII-40). For a discussion of the translation of 'zu Unrecht', see K, p. 306, n. 46.

Until more is said about how exactly the appeal to communal practice makes the middle course available, this is only a programme for a solution to Wittgenstein's problem. But even if we were at a loss as to how he might have thought the programme could be executed (and I shall suggest that we need not be: see Sections 10 and 11 below), this would be no ground for ignoring the clear textual evidence that the programme is Wittgenstein's own.

8

What I have claimed might be put like this: Wittgenstein's point is that we have to situate our conception of meaning and understanding within a framework of communal practices. Kripke's reading credits Wittgenstein with the thesis that the notion of meaning something by one's words is 'inapplicable to a single person considered in isolation' (K, p. 277). The upshot is similar, then; and it cannot be denied that the insistence on publicity in Kripke's reading corresponds broadly with a Wittgensteinian thought. But it makes a difference how we conceive the requirement of publicity to emerge.

In my reading, it emerges as a condition for the intelligibility of rejecting a premiss—the assimilation of understanding to interpretation—that would present us with an intolerable dilemma. So there are three positions in play: the two horns of the dilemma, and the community-oriented conception of meaning that enables us to decline the choice. Kripke conflates two of these, equating the paradox of *PI* §201—the first horn of the dilemma—with Wittgenstein's conclusion; only so can he take it that when Wittgenstein objects to the 'superlative fact' of *PI* §192, he is embracing the paradox of §201.[29] But this is quite wrong. The paradox that Wittgenstein formulates at §201 is not, as Kripke supposes, the mere 'paradox' that if we consider an individual in isolation, we do not have the means to make sense of the notion of meaning (something we might hope to disarm by appealing to the idea of a linguistic community). It is the genuine and devastating paradox that meaning is an illusion. Focusing on the individual in isolation from any linguistic community is not the way we fall into this abyss; it is, rather, an aspect of the way we struggle not to, so long as we retain the assumption that generates the dilemma. (See Section 4 above, on the idiolectic implications of the second horn.) The fundamental trouble is that Kripke makes

[29] See n. 18 above.

nothing of Wittgenstein's concern to reject the assimilation of understanding to interpretation; and the nemesis of this oversight is the unconvincingness (see Section 3 above) of the 'sceptical solution' on which Kripke's Wittgenstein must rely.

9

Kripke suggests (K, p. 239) that, in the light of *PI* §202, we should take it that the essentials of the Private Language Argument are contained in the general discussion of rule-following, rather than in the section of the *Investigations* that begins at §243, where it has been more usual to look. I cannot accept Kripke's view that the Private Language Argument is a corollary of the 'sceptical solution'; but his structural proposal can be detached from that.

Kripke remarks (K, pp. 277–8) that the lesson of Wittgenstein's reflections on rule-following is particularly counter-intuitive in two areas: mathematics and talk of 'inner' facts. This remark is still true after we have corrected Kripke's account of what the lesson is. In the case of mathematics, the difficulty is that we tend to construe the phenomenology of proof as a matter of glimpses of the super-rigid machinery in operation. In the case of talk of 'inner' facts, the difficulty lies in the temptation to suppose that one knows what one means from one's own case (*PI* §347). How can one's linguistic community have any bearing on the matter—beyond its control over the circumstances in which one gave oneself one's private ostensive definitions? Kripke's illuminating suggestion is that the passages usually regarded as containing the Private Language Argument are not rightly so regarded; the argument is essentially complete by *PI* §202, and the familiar passages (§§258, 265, 293, and so forth) are attempts to dissipate this inclination to cite talk of 'inner facts' as a counter-example to its conclusion.

This implies that whether those familiar passages carry conviction is, in a sense, irrelevant to the cogency of Wittgenstein's argument. If the inclination to regard talk of 'inner' facts as a counter-example persists through them, that by itself cuts no ice. And we are now in a position to see what would be needed in order to undermine the argument. One would need to show either that one or the other of the horns of the dilemma can be comfortably occupied, or that it is not the case that the assimilation of understanding to interpretation, which poses the dilemma, can be resisted only by locating meaning in a framework of communal practices.

If the target of Wittgenstein's reflections is the assimilation of understanding to interpretation, we should expect the areas where his conclusion is peculiarly counter-intuitive to be areas where we are strongly inclined to be comfortable with that assimilation. In the mathematical case, we are particularly prone to the assimilation because—as I remarked above—we are especially inclined to accept its natural accompaniment, the picture of the super-rigid machine. What about talk of 'inner' facts? We are strongly tempted, in this context, to think that there could be a private grasp of a concept—something by which, for all its privacy, it would make sense to think of judgements and utterances as constrained. What Wittgenstein's argument, as I read it, requires is the diagnosis that we are here toying with the picture of an interpretation (placed by us on a private ostensive definition)—that it is only so that we can contrive to conceive the matter in terms of concepts and judgements at all. It is true that this pictured interpretation does not readily succumb to the softening effect of the sceptical reasoning—'one interpretation after another, as if each one contented us at least for a moment, until we thought of yet another standing behind it' (*PI* §201). We imagine that in this case we can picture an interpretation that stays hard—one that comprehensively bridges the gap between the private ostensive definition and the judgements that we picture it as dictating. But there cannot be exceptions to the thesis that no interpretation can bridge the gap between the acquisition of a concept and its subsequent employment. It is this, I think, that Wittgenstein is trying to make vivid for us in the battery of passages of which this might stand as an epitome:

> Always get rid of the idea of the private object in this way: assume that it constantly changes, but that you do not notice the change because your memory constantly deceives you. (*PI* p. 207.)[30]

The idea that a private interpretation can be immune to the softening effect must be an illusion. If we conceive such an interpretation as comprehensively filling the gap, whatever the gap turns out to be, we deprive of all substance the hardness that we picture it as having.

It may be tempting to locate a weakness, in the argument I attribute to Wittgenstein, in the claim that we can steer between Scylla and Charybdis only by appealing to the practice of a community. If it is the notion of a practice that does the work, can we not form a conception of

[30] See e.g. *PI* §§258, 265, 270. See Anthony Kenny, 'The Verification Principle and the Private Language Argument', in O. R. Jones (ed.), *The Private Language Argument* (London: Macmillan, 1971), 204–28.

the practice of an individual that would do the trick?[31] But if one is tempted by this thought, one must search one's conscience to be sure that what one has in mind is not really, after all, the picture of a private interpretation; in which case one is not, after all, steering between Scylla and Charybdis, but resigning oneself to Scylla, leaving oneself fully vulnerable to the line of argument that I have just sketched.[32]

10

Wright's reading of Wittgenstein hinges on this conditional: if possession of a concept were correctly conceived as grasp of a (ratification-independent) pattern, then there would be no knowing for sure how someone else understands an expression. This conditional underlies Wright's conviction that, when we entertain the 'pattern' idea,

> the kind of reflective grasp of meaning appealed to is essentially *idiolectic*—it is a matter of each of us discerning the character of his own understanding of expressions. There is no temptation to claim a reflective knowledge of features of *others'* understanding of a particular expression—except against the background of the hypothesis that it coincides with one's own.[33]

We can summarize Wright's reading by saying that he takes Wittgenstein to propound a *modus tollens* argument with the conditional as major premiss. Thus: the idea of knowledge of idiolectic meaning is an illusion; therefore possession of a concept cannot be correctly conceived as grasp of a (ratification-independent) pattern.

The basis of this argument is, as Wright points out, 'the fundamental anti-realist thesis that we have understanding only of concepts of which we can distinctively manifest our understanding' (W, p. 221). Wright would ground both premisses of the *modus tollens* argument on 'anti-realism'. The justification for the minor premiss (see Section 2 above) is that the picture of an idiolectic rule makes no room for a distinction between actually conforming and merely having the impression that one is conforming. In Wright's reading the thought here is an 'anti-realist' one: that in an idiolectic context one could not distinctively manifest—

[31] Simon Blackburn presses what is in effect this question, in the unpublished paper mentioned in the first footnote to this essay. See Section 11 below.

[32] In this section I have aimed to describe only the *structure* of the Private Language Argument. A fuller account of how it works would require, in addition, discharging the unfinished business noted at the end of Section 7 above. See especially Section 11 below.

[33] W, p. 354. A footnote adds: 'Or with one's understanding of another specified expression.'

not even with a manifestation to oneself—a difference in one's understanding of 'I am actually conforming' and 'I have the impression of conforming'.[34] What underlies the major premiss—the conditional—is the 'anti-realist' conception of what it is to manifest understanding to others.

According to that conception, the behaviour that counts as manifesting understanding to others must be characterizable, in such a way as to display its status as such a manifestation, without benefit of a command of the language in question. Without that proviso, the 'manifestation challenge' that 'anti-realists' direct against the truth-conditional conception of meaning would be trivialized.[35] The challenge would hold no fears for the truth-conditional conception if one were allowed to count as satisfying the requirement of manifestation by such behaviour as saying—manifestly, at least to someone who understands the language one is speaking—that such and such is the case. So the distinctive manifestations allowed by 'anti-realism' consist, rather, in such behaviour as assenting to a sentence in such and such circumstances.[36]

Now what—besides itself—could be fully manifested by a piece of behaviour, or a series of pieces of behaviour, described in accordance with the 'anti-realist' requirement?[37] Perhaps the behaviour would license us to attribute a disposition; but how can we extrapolate to a determinate conception of what the disposition is a disposition to do? Our characterization of the manifesting behaviour is not allowed to exploit understanding of the language in question; so even if, in our innocence, we start out by conceiving that as grasp of 'a network of determinate patterns' (W, p. 220), we are debarred from extrapolating along the pathways of the network. It seems clear that within the rules of

[34] This is how Wright thinks the Private Language Argument is to be understood. Note that the requirement of manifestation is not initially imposed, in this line of thought, as a requirement of *public* manifestation: we are brought to see that public manifestation is what is required in consequence of an independent (non-question-begging) critique of the idea of idiolectic understanding. On the structure of Wright's reading, see Section 14 below.

[35] For the terminology 'manifestation challenge', see Wright, 'Realism, Truth-Value Links, Other Minds and the Past', *Ratio*, 22 (1980), 112–32, at pp. 112–13. For the substance of the challenge, see e.g. Michael Dummett, *Frege: Philosophy of Language* (London: Duckworth, 1973), 467.

[36] It is actually an illusion to think that this kind of characterization of behaviour conforms to the 'anti-realist' requirement: see my 'Anti-Realism and the Epistemology of Understanding', in Herman Parret and Jacques Bouveresse (eds.), *Meaning and Understanding* (Berlin: De Gruyter, 1981), 225–48, at pp. 244–6. But in the course of arguing, as I am, that the programme is misconceived in principle, there is no point of jibbing at the details of its purported execution.

[37] For 'fully', see Dummett, *Frege: Philosophy of Language*, 467.

this game any extrapolation could only be inductive, which means that if we accept the requirement that understanding be fully manifested in behaviour, no extrapolation is licensed at all. The upshot is this: the 'anti-realist' requirement of manifestation precludes any conception of understanding as grasp of a network of patterns. And this is precisely the conclusion that Wright draws.[38]

The obstacle to accepting this argument is the normative character of the notion of meaning. As I have granted, Wright aims to accommodate that: he would insist that his conclusion is not that concepts have no normative status, but that the patterns they dictate are not independent of our ratification. But the trouble is (see Sections 5 and 7) that the denial of ratification-independence, by Wright's own insistence, yields a picture of the relation between the communal language and the world in which norms are obliterated. And once we have this picture, it seems impossible simply to retain alongside it a different picture, in which the openness of an individual to correction by his fellows means that he is subject to norms. The first picture irresistibly claims primacy, leaving our openness to correction by our fellows looking like, at best, an explanation of our propensity to the illusion that we are subject to norms. If this is correct, it turns Wright's argument on its head: a condition for the possibility of finding real application for the notion of meaning at all is that we reject 'anti-realism'.

I think this transcendental argument against 'anti-realism' is fully cogent. But it is perhaps unlikely to carry conviction unless supplemented with a satisfying account of how 'anti-realism' goes wrong. (Providing this supplementation will help to discharge the unfinished business noted at the end of Section 7.)

[38] At least in W. Contrast 'Strawson on Anti-Realism', *Synthese*, 40 (1979), 283–99, at p. 294: '. . . suppose [someone] has this knowledge: of every state of affairs criterially warranting the assertion, or denial, of "John is in pain", he knows in a practical sense both that it has that status and under what circumstances it would be brought out that its status was merely criterial; that is, he knows the "overturn-conditions" of any situation criterially warranting the assertion, or denial, of "John is in pain". *No doubt we could not know for sure* that someone had this knowledge; but the stronger our grounds for thinking that he did, the more baffling would be the allegation that he did not grasp the assertoric content of "John is in pain".' (My emphasis.) Here Wright contemplates maintaining a version (formulated in terms of criteria) of the idea that understanding is grasp of a pattern of use, and accordingly opts—as his overall position indeed requires—for the other horn of this dilemma: the thesis, namely, that one cannot have certain knowledge of the character of someone else's understanding. What is remarkable is Wright's insouciance about this move: it openly flouts the fundamental motivation of 'anti-realism', which is what Wright is supposed to be defending against Strawson. It seems clear that the contrasting position of W is the only one an 'anti-realist' can consistently occupy.

11

According to 'anti-realism', people's sharing a language is constituted by appropriate correspondences in their dispositions to linguistic behaviour, as characterized without drawing on command of the language, and hence not in terms of the contents of their utterances. The motivation for this thesis is admirable: a recoil from the idea that assigning a meaning to an utterance by a speaker of one's language is forming a hypothesis about something concealed behind the surface of his linguistic behaviour. But there are two possible directions in which this recoil might move one. One—the 'anti-realist' direction—is to retain the conception of the surface that makes the idea natural, and resolutely attempt to locate meaning on the surface, so conceived. That this attempt fails is the conclusion of the transcendental argument. The supplementation that the argument needs is to point out the availability of the alternative direction: namely, to reject the conception of the surface that 'anti-realism' shares with the position it recoils from. According to this different view, the outward aspect of linguistic behaviour—what a speaker makes available to others—must be characterized in terms of the contents of utterances (the thoughts they express). Of course such an outward aspect cannot be conceived as made available to just anyone; command of the language is needed in order to put one in direct cognitive contact with that in which someone's meaning consists.[39] (This might seem to represent command of the language as a mysterious sort of X-ray vision; but only in the context of the rejected conception of the surface.)

Wittgenstein warns us not to try to dig below 'bedrock'. But it is difficult, in reading him, to avoid acquiring a sense of what, as it were, lies down there: a web of facts about behaviour and 'inner' episodes, describable without using the notion of meaning. One is likely to be struck by the sheer contingency of the resemblances between individuals on which, in this vision, the possibility of meaning seems to depend, and hence impressed by an apparent precariousness in our making sense of one another.[40] There is an authentic insight here, but one that is easily distorted; correcting the distortion will help to bring out what is wrong with the 'anti-realist' construal of Wittgenstein.

The distorted version of the insight can be put as a dilemma, on these

[39] See my 'Anti-Realism and the Epistemology of Understanding', esp. pp. 239–44.

[40] See K, p. 290; cf. Stanley Cavell, *Must we Mean What We Say?* (New York: Scribner, 1969), 52: and pp. 145–54 of my 'Non-Cognitivism and Rule-Following', in Holtzman and Leich, *To Follow a Rule*, 141–62.

lines. Suppose that, in claiming a 'reflective knowledge' of the principle of application of some expression, I claim to speak for others as well as myself. In that case my claim (even if restricted to a definitely specified other: say my interlocutor in a particular conversation) is indefinitely vulnerable to the possibility of an unfavourable future. Below 'bedrock' there is nothing but contingency; so at any time in the future my interlocutor's use of the expression in question may simply stop conforming to the pattern that I expect. And that would retrospectively undermine my present claim to be able to vouch for the character of his understanding. So I can claim to know his pattern now only 'against the background of the hypothesis that it coincides with [my] own' (W, p. 354). If, then, we retain the conception of understanding as grasp of patterns, the feeling of precariousness becomes the idea that what we think of as a shared language is at best a set of corresponding idiolects, with our grounds for believing in the correspondence no better than inductive. The only alternative—the other horn of the dilemma—is, with Wright, to give up the conception of understanding as grasp of (ratification-independent) patterns. This turns the feeling of precariousness into the idea that I cannot know for sure that my interlocutor and I will continue to march in step. But on this horn my present claim to understand him is not undermined by that concession: my understanding him now is a matter of our being in step now, and does not require a shared pattern extending into the future.

What is wrong with this, in Wittgensteinian terms, is that it conflates propositions at (or above) 'bedrock' with propositions about the contingencies that lie below. (See, for instance, *RFM* VI-49.) Its key thought is that, if I claim to know someone else's pattern, I bind myself to a prediction of the uses of language that he will make in various possible future circumstances, with these uses characterized in sub-'bedrock' terms. (That is why coming to see the contingency of the resemblances, at this level, on which meaning rests is supposed to induce appreciation that knowledge of another person's pattern could at best be inductive.) But when I claim understanding of someone else, and construe this as knowledge of the patterns to which his present utterance owes allegiance, what I claim to know is not that in such and such circumstances he will do so and so, but rather at most that that is what he will do if he sticks to his patterns.[41] And that is not a prediction at all. (Compare *RFM* VI-15.)

[41] Even this is too much. It passes muster where the 'pattern' idea is least metaphorical, namely in the case of continuation of a series; but in the general case, the idea of a corpus of

It is true that a certain disorderliness below 'bedrock' would undermine the applicability of the notion of rule-following. So the underlying contingencies bear an intimate relation to the notion of rule-following—a relation that Wittgenstein tries to capture by saying 'It is as if we had hardened the empirical proposition into a rule' (*RFM* VI-22). But recognizing the intimate relation must not be allowed to obscure the difference of levels.[42] If we respect the difference of levels, what we make of the feeling of precariousness will be as follows. When I understand another person, I know the rules he is going by. My right to claim to understand him is precarious, in that nothing but a tissue of contingencies stands in the way of my losing it. But to envisage its loss is not necessarily to envisage its turning out that I never had the right at all. The difference of levels suffices to drive a wedge between these; contrast the second horn of the above dilemma, on which inserting the wedge requires abandonment of the idea that mutual understanding is mutual knowledge of shared commitments.[43]

'Anti-realists' hold that initiation into a common language consists in acquisition of linguistic propensities describable without use of the notion of meaning. They thereby perpetrate exactly the conflation of levels against which Wittgenstein warns; someone's following a rule, according to 'anti-realism', is constituted by the obtaining of resemblances, describable in sub-'bedrock' terms, between his behaviour and that of his fellows. Not that 'anti-realists' would put it like that: it is

determinate predictions to which a claim of present understanding would commit one is absurd. (See n. 3 above.) The point I am making here is a version of one that Rush Rhees makes, in terms of a distinction between the general practice of linguistic behaviour and the following of rules, at pp. 55–56 of 'Can there be a Private Language?', in his *Discussions of Wittgenstein* (London: Routledge and Kegan Paul, 1970), 55–70. It disarms, as an objection to Wittgenstein, the insightful remarks of Jerry A. Fodor, *The Language of Thought* (Hassocks: Harvester, 1975), 71–2.

[42] The difference of levels is the subject of Wittgenstein's remarks about 'the limits of empiricism': *RFM* III-71, VII-17, VII-21. (The source of the phrase is Russell's paper of that name, in *Proceedings of the Aristotelian Society*, 36 (1935–6), 131–50.) See W, p. 220. I think the point of the remarks is, very roughly, that empiricism can deal only with what is below 'bedrock': the limits of empiricism (which 'are not assumptions unguaranteed, or intuitively known to be correct: they are ways in which we make comparisons and in which we act': *RFM* VII-21—cf. *On Certainty* §204, quoted in Section 7 above) live above it (outside its reach), at 'bedrock' level. Wright, by contrast, seems to interpret the passages as if Wittgenstein's view were that for all its limits empiricism contained the truth.

[43] Christopher Peacocke, at p. 88 of 'Rule-Following: the Nature of Wittgenstein's Arguments', in Holtzman and Leich (eds.), *To Follow a Rule*, 72–95, implies that statements about rule-following *supervene*, in Wittgenstein's view, on sub-'bedrock' statements. There may be an acceptable interpretation of this; but on the most natural interpretation, it would make statements about rule-following vulnerable to future loss of mutual intelligibility in just the way I am objecting to.

another way of making the same point to say that they locate 'bedrock' lower than it is—not accommodating the fact that 'following according to the rule is FUNDAMENTAL to our language-game' (*RFM* VI-28; see Section 7 above). If, by contrast, we satisfy the motivation of 'anti-realism' in the different way that I distinguished above, then we refuse to countenance sub-'bedrock' (meaning-free) characterizations of what meaning something by one's words consists in, and thus respect Wittgenstein's distinction of levels.

We make possible, moreover, a radically different conception of what it is to belong to a linguistic community. 'Anti-realists' picture a community as a collection of individuals presenting to one another exteriors that match in certain respects. They hope to humanize this bleak picture by claiming that what meaning consists in lies on those exteriors as they conceive them. But the transcendental argument reveals this hope as vain. A related thought is this: if regularities in the verbal behaviour of an isolated individual, described in norm-free terms, do not add up to meaning, it is quite obscure how it could somehow make all the difference if there are several individuals with matching regularities.[44] The picture of a linguistic community degenerates, then, under 'anti-realist' assumptions, into a picture of a mere aggregate of individuals whom we have no convincing reason not to conceive as opaque to one another. If, on the other hand, we reject the 'anti-realist' restriction on what counts as manifesting one's understanding, we entitle ourselves to this thought: shared membership in a linguistic community is not just a matter of matching in aspects of an exterior that we present to anyone whatever, but equips us to make our minds available to one another, by confronting one another with a different exterior from that which we present to outsiders.

Wittgenstein's problem was to explain how understanding can be other than interpretation (see Section 7 above). This non-'anti-realist' concep-

[44] Simon Blackburn, at p. 183 of 'Rule-Following and Moral Realism', writes: '. . . we can become gripped by what I call a *wooden* picture of the use of language, according to which the only fact of the matter is that in certain situations people use words, perhaps with various feelings like "that fits", and so on. This wooden picture makes no room for the further fact that in applying or withholding a word people may be conforming to a pre-existent rule. But just because of this, it seems to make no room for the idea that in using their words they are expressing judgements. Wittgenstein must have felt that publicity, the fact that others do the same, was the magic ingredient turning the wooden picture into the full one. It is most obscure to me that it fills this role: a lot of wooden persons with propensities to make noises is just more of whatever one of them is.' It will be apparent that I have a great deal of sympathy with this complaint. Where I believe Blackburn goes wrong is in thinking that it tells against Wittgenstein himself, as opposed to the position that Wittgenstein has been saddled with by a certain set of interpreters (among whom I did not intend to enrol myself in my 'Non-Cognitivism and Rule-Following', the paper to which Blackburn is responding).

tion of a linguistic community gives us a genuine right to the following answer: shared command of a language equips us to know one another's meaning without needing to arrive at that knowledge by interpretation, because it equips us to hear someone else's meaning in his words. 'Anti-realists' would claim this right too, but the claim is rendered void by the merely additive upshot of their picture of what it is to share a language. In the different picture I have described, the response to Wittgenstein's problem works because a linguistic community is conceived as bound together, not by a match in mere externals (facts accessible to just anyone), but by a capacity for a meeting of minds.

When we had no more than an abstract characterization of Wittgenstein's response, as an appeal to the notion of communal practice, there seemed to be justice in this query: if the concept of a communal practice can magic meaning into our picture, should not this power be credited to the concept of a practice as such—so that the practice of an individual might serve just as well? (See Section 7 above.) But if Wittgenstein's position is the one I have described in this section, it is precisely the notion of a communal practice that is needed, and not some notion that could equally be applied outside the context of a community. The essential point is the way in which one person can know another's meaning without interpretation. Contrary to Wright's reading, it is only because we *can* have what Wright calls 'a reflective knowledge of features of *others*' understanding of a particular expression' (W, p. 354) that meaning is possible at all.[45]

12

Wittgenstein's reflections on rule-following attack a certain familiar picture of facts and truth, which I shall formulate like this. A genuine fact must be a matter of the way things are in themselves, utterly independently of us. So a genuinely true judgement must be, at least potentially, an exercise of pure thought; if human nature is necessarily implicated in the very formation of the judgement, that precludes our thinking of the corresponding fact as properly independent of us, and hence as a proper fact at all.[46]

[45] If I am right to suppose that any merely aggregative conception of a linguistic community falsifies Wittgenstein, then it seems that the parallel that Kripke draws with Hume's discussion of causation (independently proposed by Blackburn, 'Rule-Following and Moral Realism', 182–83) is misconceived. Wittgenstein's picture of language contains no conception of the individual such as would correspond to the individual cause-effect pair, related only by contiguity and succession, in Hume's picture of causation.

[46] The later Wittgenstein may have (perhaps unjustly) found a form of this picture in the

We can find this picture of genuine truth compelling only if we either forget that truth-bearers are such only because they are meaningful, or suppose that meanings take care of themselves, needing, as it were, no help from us. This latter supposition is the one that is relevant to our concerns. If we make it, we can let the judging subject, in our picture of true judgement, shrink to a locus of pure thought, while the fact that judging is a human activity fades into insignificance.

Now Wittgenstein's reflections on rule-following undermine this picture by undermining the supposition that meanings take care of themselves. A particular performance, 'inner' or overt, can be an application of a concept—a judgement or a meaningful utterance—only if it owes allegiance to constraints that the concept imposes. And being governed by such constraints is not being led, in some occult way, by an autonomous meaning (the super-rigid machinery), but acting within a communal custom. The upshot is that if something enters into being a participant in the relevant customs, it enters equally into being capable of making any judgements at all. We have to give up that picture of genuine truth, in which the maker of a true judgement can shrink to a point of pure thought, abstracted from anything that might make him distinctively and recognizably one of us.

It seems right to regard that familiar picture as a kind of realism. It takes meaning to be wholly autonomous (one is tempted to say 'out there'); this is reminiscent of realism as the term is used in the old debate about universals. And it embraces an extreme form of the thesis that the facts are not up to us; this invites the label 'realism' understood in a way characteristic of more recent debates. But if we allow ourselves to describe the recoil from the familiar picture as a recoil from realism, there are two points that we must be careful not to let this obscure.

First: the recoil has nothing to do with rejection of the truth-conditional conception of meaning, properly understood. That conception has no need to camouflage the fact that truth conditions are necessarily given by us, in a language that we understand. When we say '"Diamonds are hard" is true if and only if diamonds are hard', we are just as much involved on the right-hand side as the reflections on rule-following tell us we are. There is a standing temptation to miss this obvious truth, and to suppose that the right-hand side somehow presents us with a possible fact, pictured as an unconceptualized configuration of

Tractatus. On the relation between the later work and the *Tractatus*, see Peter Winch, 'Introduction: The Unity of Wittgenstein's Philosophy', in Peter Winch (ed.), *Studies in the Philosophy of Wittgenstein* (London: Routledge & Kegan Paul, 1969), 1–19, esp. the very illuminating discussion at pp. 9–15.

things in themselves. But we can find the connection between meaning and truth illuminating without succumbing to that temptation.

Second: the recoil is from an extreme form of the thesis that the facts are not up to us, not from that thesis in any form whatever. What Wittgenstein's polemic against the picture of the super-rigid machine makes untenable is the thesis that possession of a concept is grasp of a pattern of application that extends *of itself* to new cases. (See Section 4 above.) In Wright's reading, that is the same as saying that it deprives us of the conception of grasp of ratification-independent patterns. But rejection of ratification-independence obliterates meaning altogether (see Sections 5, 7, 10 above). In effect, the transcendental argument shows that there *must* be a middle position. Understanding is grasp of patterns that extend to new cases independently of our ratification, as required for meaning to be other than an illusion (and—not incidentally—for the intuitive notion of objectivity to have a use); but the constraints imposed by our concepts do not have the platonistic autonomy with which they are credited in the picture of the super-rigid machinery.

As before (compare Section 11 above), what obscures the possibility of this position is the 'anti-realist' attempt to get below 'bedrock'. Wright suggests (W, pp. 217–20) that the emergence of a consensus on whether, say, to call some newly encountered object 'yellow' is subject to no norms. That is indeed how it seems if we allow ourselves to picture the communal language in terms of sub-'bedrock' resemblances in behaviour and phenomenology. But if we respect Wittgenstein's injunction not to dig below the ground, we must say that the community 'goes right or wrong' (compare W, p. 220) according to whether the object in question is, or is not, *yellow*; and nothing can make its being yellow, or not, dependent on our ratification of the judgement that that is how things are. In Wittgenstein's eyes, as I read him, Wright's claim that 'for the community itself there is no authority, so no standard to meet' (W, p. 220) can be, at very best, an attempt to say something that cannot be said but only shown. It may have some merit, conceived in that light; but attributing it to Wittgenstein as a doctrine can yield only distortion.

Wittgenstein writes, at *RFM* II-61:

> Finitism and behaviourism are quite similar trends. Both say, but surely, all we have here is . . . . Both deny the existence of something, both with a view to escaping from a confusion.[47]

[47] Kripke discusses this passage at K, pp. 293–294; but I believe his attribution to Wittgenstein of the 'sceptical paradox' and the 'sceptical solution' prevents him from fully appreciating its point.

The point about finitism is this. It recoils, rightly, from the mythology of the super-rigid machinery—the patterns that extend of themselves, without limit, beyond any point we take them to. But it equates this recoil with rejecting any conception of patterns that extend, without limit, beyond any such point. This is like the behaviourist idea that in order to escape from the confused idea of the mental as essentially concealed from others behind behaviour, we have to reject the mental altogether. The idealism that Wright reads into Wittgenstein seems to be another similar trend. (Clearly the remark does not applaud the trends it discusses.)

13

In this section I want to mention two sets of passages in Wittgenstein of which we are now placed to make better sense than Wright can.

First: in Wright's reading, the 'pattern' idea is inextricably connected with the picture of idiolectic understanding. But this does not seem to be how Wittgenstein sees things. Wittgenstein does not scruple to say that a series 'is defined . . . by the training in proceeding according to the rule' (*RFM* VI-16). And at *Zettel* §308 he writes:

> Instead of "and so on" he might have said: "Now you know what I mean." And his explanation would simply be a *definition* of the expression "following the rule + 1" . . .

Again, *PI* §208 and the remarks that follow it contain a sustained attack on the idea that successfully putting someone through the sort of training that is meant to 'point beyond' the examples given (see §208) is getting him 'to *guess* the essential thing' (*PI* §210). For Wright, when these passages reject the picture of a leap to a personal understanding, they should be *eo ipso* rejecting the 'pattern' idea. But Wittgenstein combines criticism of the 'leap' picture with conceding (§209) how natural it is to think of our understanding as reaching beyond all the examples given. (Wright would construe this concession in terms of his purged version of the 'pattern' idea. But we can make sense of what Wittgenstein says without saddling him with the problems generated by denial of ratification-independence.)

Second: Wittgenstein sometimes (for instance at *PI* §151) discusses the idea that one can grasp the principle of a series, or a meaning, 'in a flash'. Wright suggests (W, pp. 30–1) that the idea of this 'flash' can be nothing but the idea of a leap to a purely personal understanding. But I see no reason to accept that Wittgenstein intends this identification. In fact, the suggestion casts a gratuitous slur on his phenomenological perceptiveness. The idea that the meaning of an expression can be present in an instant is just as tempting about someone else's meaning as it is about one's own; and Wittgenstein is perfectly aware of this:

> When someone says the word "cube" to me, for example, I know what it means. But can the whole *use* of the word come before my mind, when I *understand* it in that way? (*PI* §139; cf. §138.)

Wright's view must be that the intended answer to this question is 'No'—that Wittgenstein intends to show up as an illusion the idea that one can grasp someone else's pattern in a flash. But the only illusion that Wright explains to us in this neighbourhood is the illusion of supposing that one could have an idiolectic grasp of a pattern. So Wright's Wittgenstein owes us something for which we search the writings of the actual Wittgenstein in vain: an explanation of how it is that we not only fall into that illusion but misconceive its character—mistaking what is in fact the supposition that we can guess at someone else's pattern for (what seems on the face of it very different) the supposition that we can hear it in his utterances.

We are now placed to see that this latter supposition is not, in Wittgenstein's view, an illusion at all. 'Grasping the whole use in a flash' is not to be dismissed as expressing an incorrigibly confused picture—the picture of a leap to an idiolectic understanding—but to be carefully understood in the light of the thesis that there is a way of grasping a rule which is not an interpretation. In that light, we can see that there is nothing wrong with the idea that one can grasp in a flash the principle of a series one is being taught; and equally that there is nothing wrong with the idea that one can hear someone else's meaning in his words. The 'interpretation' prejudice insidiously tempts us to put a fantastic mythological construction on these conceptions; the right response to that is not to abandon the conceptions but to exorcize the 'interpretation' prejudice and so return them to sobriety. ('Really the only thing wrong with what you say is the expression "in a queer way"': *PI* §195.)

At *PI* §534, Wittgenstein writes:

> *Hearing* a word in a particular sense. How queer that there should be such a thing!
>
> Phrased *like this*, emphasized like this, heard in this way, this sentence is the first of a series in which a transition is made to *these* sentences, pictures, actions.
>
> ((A multitude of familiar paths leads off from these words in every direction.))[48]

[48] The last sentence is quoted from *PI* §525. A related passage is *PI* IIxi: the connection between the topics of seeing an aspect and 'experiencing the meaning of a word' is drawn explicitly at pp. 214, 215.

What are these 'familiar paths'? Presumably, for instance, continuations of the conversation that would make sense: not, then, 'patterns' in precisely the sense with which we have been concerned (which would be, as these paths would not, cases of 'going on doing the same thing'), but they raise similar issues. Suppose that, in describing a series of utterances that in fact constitutes an intelligible conversation, we conform to the 'anti-realist' account of how meaning must be manifested. We shall have to describe each member of the series without drawing on command of the language in question. Such a description will blot out the relations of meaning between the members of the series, in virtue of which it constitutes an intelligible conversation; what is left will be, at best, a path that one could trace out inductively (whether predicting or retrodicting).[49] Wright's demonstration that 'anti-realism' cannot countenance ratification-independent patterns should work for these 'familiar paths' too. An 'anti-realist' cannot extrapolate, from what is done in his presence on an occasion, along paths marked out by meaning; and inductive extrapolation is against the rule that we must restrict ourselves to what is fully manifested in linguistic behaviour. It is obscure to me what interpretation of the passage I have quoted is available to Wright. What seems to be the case is that 'anti-realism', by, in effect, looking for 'bedrock' lower than it is, blocks off the obvious and surely correct reading: that hearing a word in one sense rather than another is hearing it in one position rather than another in the network of possible patterns of making sense that we learn to find ourselves in when we acquire mastery of a language.

14

We can centre the issue between Wright's reading and mine on this question: how does Wittgenstein's insistence on publicity emerge? In my reading, the answer is this: it emerges as a condition of the possibility of rejecting the assimilation of understanding to interpretation, which poses an intolerable dilemma. In Wright's reading, the answer is this: it

[49] At pp. 130–31 of 'What is a Theory of Meaning? (II)', Dummett writes: 'We do not expect, nor should we want, to achieve a deterministic theory of meaning for a language, even one which is deterministic only in principle: we should not expect to be able to give a theory from which, together with all other relevant conditions (the physical environment of a speaker, the utterances of other speakers, etc.), we could predict the exact utterances of any one speaker, any more than, by a study of the rules and strategy of a game, we expect to be able to predict actual play.' But in the context of the 'anti-realist' restriction, all that this can mean is that we must content ourselves with weaker relations of the same general kind (inductively traceable, not meaning-dependent) as those that would be involved in a theory of the deterministic sort we are to renounce.

emerges as the only alternative left, after the notion of idiolectic understanding has been scotched by a self-contained argument that is epitomized by this passage (*PI* §258): 'One would like to say: whatever is going to seem right to me is right. And that only means that here we can't talk about "right".' Wright takes the thought here to be an 'anti-realist' one, to the effect that the distinction between being right and seeming right is shown to be empty, in the idiolectic case, by the impossibility of manifesting a grasp of it, even to oneself. (See Section 10 above.) Given this, I suppose Wright takes it that sheer consistency requires construing the appeal to the community, shown to be obligatory by virtue of being the only remaining possibility, in an 'anti-realist' way.

Now it is true that the idiolectic conception of understanding is a corollary of the second horn of the dilemma. (See Section 4 above.) So my reading need not exclude a self-contained argument against that idea, constituting part of the demonstration that the dilemma is intolerable. On such a view, the insistence on publicity would emerge twice over: first as a direct implication of the self-contained argument, and second, indirectly, as required by the rejection of the dilemma. In fact I think this complexity is unnecessary. Wittgenstein has plenty to say against the second horn of the dilemma—the picture of the super-rigid machine—without needing, for his case against it and therefore against accepting the dilemma, the envisaged self-contained argument against this corollary. And I have explained (in Section 9 above) how passages like the one I quoted above from *PI* §258, which Wright takes as formulations of the self-contained argument, are intelligible in the context of the second, indirect route to the requirement of publicity. But the real flaw in Wright's reading, in my view, is not that it countenances the first route, but that it omits the second. Like Kripke (see Section 8 above), Wright makes nothing of Wittgenstein's concern—which figures at the centre of my reading—to attack the assimilation of understanding to interpretation.

This oversight shows itself in Wright's willingness to attribute the following line of thought to Wittgenstein:

> the investigation-independent truth of statements requires that their truth is settled, autonomously and without the need of human interference, by their meanings and the character of the relevant facts. For a complex set of reasons, however, no notion of meaning can be legitimised which will play this role . . . the meaning of a statement, if it is to make the relevant autonomous contribution towards determining that statement's truth-value, cannot be thought of as fully determined by previous uses of that statement or, if it is a novel statement, by previous uses of its constituents and by its syntax; for those factors can always be reconciled with the statement's having any truth-value, no matter what the

> worldly facts are taken to be. The same goes for prior phenomenological episodes—imagery, models—in the minds of the linguistically competent. Nothing, therefore, in the previous use of the statement, or of its constituents, or in the prior streams of consciousness of competent speakers, is, if its meaning is in conjunction with the facts to determine its truth-value, sufficient to fix its meaning. So what does?[50]

This is essentially the argument that generates the paradox of *PI* §201; and it can be attributed to Wittgenstein only at the cost of ignoring, like Kripke, that section's second paragraph.

The result of the oversight is that, whereas Wittgenstein's key thought is that the dilemma must be avoided, Wright's reading leaves the dilemma unchallenged. Wittgenstein obviously attacks the second horn of the dilemma—the picture of the super-rigid machinery. The consequence of leaving the dilemma unchallenged is thus to locate Wittgenstein on its first horn—embracing the paradox of §201. This disastrous upshot does not, of course, correspond to Wright's *intentions* in his interpretation of Wittgenstein. (Contrast Kripke, who can be content to attribute acceptance of the paradox of §201 to Wittgenstein because he misses its devastating character.) Nevertheless, it is where his reading leaves us (see Sections 5, 7, 10 above): a fitting nemesis for its inattention to Wittgenstein's central concern.

The villain of the piece—what makes it impossible for Wright to accommodate Wittgenstein's insistence that understanding need not be interpretation—is the 'anti-realist' conception of our knowledge of others. (See Sections 11 and 12 above. Contrary to what, at the beginning of this section, I took Wright to suppose, the cogency of a passage like *PI* §258, against the picture of idiolectic understanding, is quite unconnected with the 'anti-realist' view of what it is to manifest understanding to others.) From Wright's reading, then, we can learn something important: that there cannot be a position that is both 'anti-realist' and genuinely hospitable to meaning, and that the construal of Wittgenstein as the source of 'anti-realism', often nowadays taken for granted, is a travesty.

[50] 'Strict Finitism', p. 250. Note also W, p. 22, where Wright identifies the second speaker in the dialogue of *RFM* I-113 ('However many rules you give me—I give a rule which justifies *my* employment of your rules') with Wittgenstein himself; and W, p. 216 (a passage quoted in Section 2 above), where it is the susceptibility of all explanations to unintended *interpretations* that is said to push us into the idea of understanding as essentially idiolectic.

NOTES ON THE CONTRIBUTORS

DONALD DAVIDSON is at the University of California, Berkeley. His principal related writings are collected together in his *Inquiries into Truth and Interpretation* (1984).

MICHAEL DUMMETT was at New College, Oxford, until his retirement in 1992. He has an emeritus position at All Souls' College, Oxford. He is particularly known for his exegesis of Frege, which has resulted in *Frege: Philosophy of Language* (1981), *The Interpretation of Frege's Philosophy* (1981), and *Frege: Philosophy of Mathematics* (1991). Many of his essays concerning questions of meaning and reference appear in his *Truth and Other Enigmas* (1978). A related book is *The Logical Basis of Metaphysics* (1991).

GARETH EVANS (1946–1980) was at University College, Oxford. His posthumously published *The Varieties of Reference* (1982) was succeeded by his *Collected Papers* (1985), many of which are concerned with meaning and reference. He also edited *Truth and Meaning* (1976) with McDowell.

GOTTLOB FREGE (1848–1925), by common consent the greatest logician of all time, taught at the University of Jena. Of the various essays in which he developed his views on sense and reference, one of the best known is 'The Thought', in Strawson (ed.), *Philosophical Logic*, in the Oxford Readings in Philosophy series. Others appear in his *Philosophical Writings* (1960).

SAUL KRIPKE is at Princeton University. His principal related writings are *Naming and Necessity* (1980) and *Wittgenstein on Rules and Private Language* (1982).

JOHN MCDOWELL is at the University of Pittsburgh. He edited *Truth and Meaning* (1976) with Evans, as well as Evans's posthumously published *The Varieties of Reference* (1982). His essays in this area appear in a range of journals and other volumes.

HILARY PUTNAM is at Harvard University. Most of his work on meaning and reference appears in the collections *Mind, Language and Reality* (1975) and *Realism and Reason* (1983), as well as in the book *Meaning and the Moral Sciences* (1978).

W. V. QUINE is a Professor Emeritus at Harvard University. His vast output in this area includes the books *Word and Object* (1960), *Philosophy of Logic* (1970), and *The Roots of Reference* (1974), as well as the collections *From a Logical Point of View* (1961) and *Ontological Relativity and Other Essays* (1969).

BERTRAND RUSSELL (1872–1970), a major philosopher of the twentieth century, was at Trinity College, Cambridge. His extensive work on questions of meaning and reference include *An Inquiry into Meaning and Truth* (1940) and 'On Denoting', which appears in his *Logic and Knowledge* (1956).

P. F. STRAWSON was at Magdalen College, Oxford, until his retirement in 1987. He now has an emeritus position at University College, Oxford. He edited *Philosophical Logic* in the Oxford Readings in Philosophy series. Most of his essays on meaning and reference are collected together in his *Logico-Linguistic Papers* (1971).

DAVID WIGGINS is at Birkbeck College, London. His essays in this area appear in a range of journals and other volumes. A related book is his *Sameness and Substance* (1980).

SELECT BIBLIOGRAPHY

1. MONOGRAPHS OF GENERAL INTEREST

Anscombe, G. E. M., *An Introduction to Wittgenstein's Tractatus* (London: Hutchinson, 1959).
Carnap, R., *Meaning and Necessity* (Chicago: University of Chicago Press, 1956).
Davies, M. K., *Meaning, Quantification, Necessity* (London: Routledge & Kegan Paul, 1981).
Dummett, M., *Frege: Philosophy of Language*, rev. edn. (London: Duckworth, 1981).
—— *The Interpretation of Frege's Philosophy* (London: Duckworth, 1981).
Evans, G., *The Varieties of Reference*, ed. J. McDowell (Oxford: Clarendon Press, 1982).
Geach, P. T., *Reference and Generality* (Ithaca, NY: Cornell University Press, 1962).
Heal, J., *Fact and Meaning* (Oxford: Blackwell, 1989), esp. helpful for Quine's views (3.3 below) and Wittgenstein's views (3.7 below).
Kripke, S. A., *Naming and Necessity* (Oxford: Blackwell, 1980), esp. helpful for 3.2 and 3.6.
Linsky, L., *Names and Descriptions* (Chicago: University of Chicago Press, 1977), esp. helpful for 3.2.
McCulloch, G., *The Game of the Name* (Oxford: Clarendon Press, 1989), esp. helpful for 3.1 and 3.2.
Putnam, H., *Meaning and the Moral Sciences* (London: Routledge & Kegan Paul, 1978), esp. helpful for 3.6.
Quine, W. V., *Word and Object* (Cambridge, Mass.: MIT Press, 1960), esp. helpful for 3.3.
—— *Philosophy of Logic* (Englewood Cliffs, NJ: Prentice-Hall, 1970).
—— *The Roots of Reference* (La Salle, Ill.: Open Court, 1974).
Russell, B., *An Inquiry into Meaning and Truth* (London: Allen & Unwin, 1940).
Salmon, N., *Reference and Essence* (Princeton, NJ: Princeton University Press, 1982), esp. helpful for 3.6.
—— *Frege's Puzzle* (Cambridge, Mass.: MIT Press, 1989), esp. helpful for 3.1.
Wittgenstein, L., *Tractatus Logico-Philosophicus*, trans. D. F. Pears and B. F. McGuinness (London: Routledge & Kegan Paul, 1961).
—— *Philosophical Investigations*, trans. G. E. M. Anscombe (Oxford: Blackwell, 1967), esp. helpful for 3.7.

2. ANTHOLOGIES OF GENERAL INTEREST

Blackburn, S. (ed.), *Meaning, Reference and Necessity* (Cambridge: Cambridge University Press, 1975).
Davidson, D., *Inquiries into Truth and Interpretation* (Oxford: Clarendon Press, 1984), esp. helpful for 3.4.

Dummett, M., *Truth and Other Enigmas* (London: Duckworth, 1978).
Evans, G., and McDowell, J. (eds.), *Truth and Meaning* (Oxford: Clarendon Press, 1976), esp. helpful for 3.4.
Guttenplan, S. (ed.), *Mind and Language* (Oxford: Clarendon Press, 1975).
Le Pore, E. (ed.), *Truth and Interpretation* (Oxford: Blackwell, 1986), esp. helpful for 3.4.
Linsky, L. (ed.), *Reference and Modality* (Oxford: Oxford University Press, 1971), esp. helpful for 3.6.
Platts, M. (ed.), *Reference, Truth and Reality* (London: Routledge & Kegan Paul, 1980).
Quine, W. V., *From a Logical Point of View* (New York: Harper & Row, 1961).
—— *Ontological Relativity and Other Essays* (New York: Columbia University Press, 1969).
Schwartz, S. (ed.), *Naming, Necessity and Natural Kinds* (Ithaca, NY: Cornell University Press, 1977), esp. helpful for 3.6.
Strawson, P. F. (ed.), *Philosophical Logic* (Oxford: Oxford University Press, 1967).
Wright, C. (ed.), *Frege: Tradition and Influence* (Oxford: Blackwell, 1986).

3. READING ON INDIVIDUAL TOPICS

This section does not include material from 1 and 2, except where attention is drawn to specific chapters or essays.

3.1. Frege's Distinction between Sense and Reference (see also 3.2)

Burge, T., 'Sinning Against Frege', *Philosophical Review*, 88 (1979).
Dummett, M., ch. 5 and appendix of *Frege: Philosophy of Language*, listed in 1.
Evans, G., 'Understanding Demonstratives', in his *Collected Papers* (Oxford: Clarendon Press, 1985).
Frege, G., 'The Thought', in P. F. Strawson (ed.), *Philosophical Logic*, listed in 2.
McDowell, J., '*De Re* Senses', in C. Wright (ed.), *Frege: Tradition and Influence*, listed in 2.
Sainsbury, R. M., 'On a Fregean Argument for the Distinction of Sense and Reference', *Analysis*, 43 (1983).
Wiggins, D., 'Frege's Problem of the Morning Star and the Evening Star', in M. Schirn (ed.), *Studies in the Philosophy of Frege*, ii (Stuttgart-Bad Canstatt: Frommann-Holzboog, 1976).
—— 'The Sense and Reference of Predicates', in C. Wright (ed.), *Frege: Tradition and Influence*, listed in 2.

3.2. Names and Descriptions (see also 3.1 and 3.6)

Donnellan, K., 'Reference and Definite Descriptions', *Philosophical Review*, 75 (1966).
—— 'Proper Names and Identifying Descriptions', in D. Davidson and G. Harman (eds.), *Semantics of Natural Language* (Dordrecht: Reidel, 1972).
Evans, G., 'Reference and Contingency', in his *Collected Papers* (Oxford: Clarendon Press, 1985).

Kaplan, D., 'What is Russell's Theory of Descriptions?', in D. F. Pears (ed.), *Bertrand Russell* (New York: Anchor Books, 1972).
Kripke, S. A., 'Semantic Reference and Speaker's Reference', in P. A. French, T. E. Uehling, and H. K. Wettstein (eds.), *Studies in the Philosophy of Language* (Minneapolis: Minnesota University Press, 1977).
—— 'A Puzzle about Belief', in A. Margalit (ed.), *Meaning and Use* (Dordrecht: D. Reidel, 1979).
Peacocke, C., 'Proper Names, Reference, and Rigid Designation', in S. Blackburn (ed.), *Meaning, Reference and Necessity*, listed in 2.
Quine, W. V., 'On What There Is', in his *From a Logical Point of View*, listed in 2.
Russell, B., 'On Denoting', *Mind*, 14 (1905).
Searle, J. R., 'Proper Names', in P. F. Strawson (ed.), *Philosophical Logic*, listed in 2.
—— 'Referential and Attributive', *Monist*, 62 (1979).

3.3. Quine's Views on the Indeterminacy of Meaning (see also 3.4)

Chomsky, N., 'Quine's Empirical Assumptions', *Synthese*, 19 (1968–9).
Davidson, D., 'Radical Interpretation', in his *Inquiries into Truth and Interpretation*, listed in 2.
—— 'The Inscrutability of Reference', in his *Inquiries into Truth and Interpretation*, listed in 2.
Dummett, M., 'The Significance of Quine's Indeterminacy Thesis', in his *Truth and Other Enigmas*, listed in 2.
Evans, G., 'Identity and Predication', in his *Collected Papers* (Oxford: Clarendon Press, 1985).
Gibson, R. F., 'Translation, Physics, and Facts of the Matter', in L. E. Hahn and P. A. Schilpp (eds.), *The Philosophy of W. V. Quine* (La Salle, Ill.: Open Court, 1986).
Grandy, R. E., 'Reference, Meaning and Belief', *Journal of Philosophy*, 70 (1973).
Grice, H. P., and Strawson, P. F., 'In Defence of a Dogma', *Philosophical Review*, 65 (1956).
Kirk, R., *Translation Determined* (Oxford: Clarendon Press, 1986).
Putnam, H., 'The Analytic and the Synthetic', in his *Mind, Language and Reality* (Cambridge: Cambridge University Press, 1975).
Quine, W. V., 'Two Dogmas of Empiricism', in his *From a Logical Point of View*, listed in 2.
—— 'Ontological Relativity', in his *Ontological Relativity and Other Essays*, listed in 2.
—— 'On the Reasons for the Indeterminacy of Translation', *Journal of Philosophy*, 67 (1970).
—— 'Indeterminacy of Translation Again', *Journal of Philosophy*, 84 (1987).
Rorty, R., 'Indeterminacy of Translation and of Truth', in *Synthese*, 23 (1971–2).
Searle, J. R., 'Indeterminacy, Empiricism and the First Person', *Journal of Philosophy*, 84 (1987).
White, M., 'The Analytic and the Synthetic: An Untenable Dualism', in L. Linsky (ed.), *Semantics and the Philosophy of Language* (Urbana, Ill.: University of Illinois Press, 1952).

Zabludowski, A., 'On Quine's Indeterminacy Doctrine', *Philosophical Review*, 98 (1989).

3.4. *Truth, Meaning, and Davidson's Project* (see also 3.3)

Dummett, M., 'What is a Theory of Meaning?', in S. Guttenplan (ed.), *Mind and Language*, listed in 2.
Field, H., 'Tarski's Theory of Truth', in M. Platts (ed.), *Reference, Truth and Reality*, listed in 2.
Lewis, D., 'Radical Interpretation', in *Synthese*, 27 (1974).
McDowell, J., 'Physicalism and Primitive Denotation', in M. Platts (ed.), *Reference, Truth and Reality*, listed in 2.
—— 'In Defence of Modesty', in B. Taylor (ed.), *Michael Dummett: Contributions to Philosophy* (Dordrecht: Nijhoff, 1987).
Platts, M., *Ways of Meaning* (London: Routledge & Kegan Paul, 1979).
Sainsbury, R. M., 'Understanding and Theories of Meaning', *Proceedings of the Aristotelian Society*, 80 (1979–80).
Strawson, P. F., 'Meaning and Truth', in his *Logico-Linguistic Papers* (London: Methuen, 1971).
Wallace, J., 'Only in the Context of a Sentence Do Words Have Any Meaning', in P. A. French, T. E. Uehling, and H. K. Wettstein (eds.), *Studies in the Philosophy of Language* (Minneapolis: Minnesota University Press, 1977).
Wright, C., *Wittgenstein on the Foundations of Mathematics* (London: Duckworth, 1979), ch. 15.
—— 'Theories of Meaning and Speakers' Knowledge', in his *Realism, Meaning and Truth* (Oxford: Blackwell, 1986).

3.5. *Realism and Anti-Realism*

Craig, E., 'Meaning, Use and Privacy', *Mind*, 91 (1982).
Dummett, M., 'Truth', in his *Truth and Other Enigmas*, listed in 2.
—— 'The Reality of the Past', in his *Truth and Other Enigmas*, listed in 2.
—— 'The Philosophical Basis of Intuitionistic Logic', in his *Truth and Other Enigmas*, listed in 2.
—— 'What is a Theory of Meaning? (II)', in G. Evans and J. McDowell (eds.), *Truth and Meaning*, listed in 2.
McDowell, J., 'Truth Conditions, Bivalence and Verificationism', in G. Evans and J. McDowell, *Truth and Meaning*, listed in 2.
—— 'On "The Reality of the Past"', in C. Hookway and P. Pettit (eds.), *Action and Interpretation* (Cambridge: Cambridge University Press, 1977).
McGinn, C., 'An A Priori Argument for Realism', *Journal of Philosophy*, 76 (1979).
Wright, C., *Realism, Meaning and Truth* (London: Duckworth, 1986).
—— 'Dummett and Revisionism', in B. Taylor (ed.), *Michael Dummett: Contributions to Philosophy* (Dordrecht: Nijhoff, 1987).

3.6. *Natural Kind Terms, Necessity, and Essentialism* (see also 3.2)

McGinn, C., 'A Note on the Essence of Natural Kinds', *Analysis*, 35 (1974–5).
Mellor, D. H., 'Natural Kinds', *British Journal for the Philosophy of Science*, 28 (1977).

Platts, M., 'Natural Kind Terms and "Rigid Designators"', *Proceedings of the Aristotelian Society*, 82 (1981–2).
Putnam, H., 'Is Semantics Possible?', in his *Mind, Language and Reality* (Cambridge: Cambridge University Press, 1975).
—— 'The Meaning of "Meaning"', in his *Mind, Language and Reality*.
Quine, W. V., 'Three Grades of Modal Involvement', in his *The Ways of Paradox and Other Essays* (New York: Random House, 1966).
Wiggins, D., *Sameness and Substance* (Oxford: Blackwell, 1980).

3.7. Wittgenstein's Views

Baker, G. P., and Hacker, P. M. S., *Scepticism, Rules and Language* (Oxford: Blackwell, 1984).
Budd, M., 'Wittgenstein on Meaning, Interpretation and Rules', *Synthese*, 58 (1984).
Dummett, M., 'Frege and Wittgenstein', in I. Block (ed.), *Perspectives on the Philosophy of Wittgenstein* (Oxford: Blackwell, 1981).
Holtzmann, S. H., and Leich, C. M. (eds.), *Wittgenstein: To Follow a Rule* (London: Routledge & Kegan Paul, 1981).
Jones, O. R. (ed.), *The Private Language Argument* (London: Macmillan, 1971).
Kripke, S. A., *Wittgenstein on Rules and Private Language* (Oxford: Blackwell, 1982).
McGinn, C., *Wittgenstein on Meaning* (Oxford: Blackwell, 1984).

INDEX OF NAMES

(not including entries in the Bibliography)